THE MOLECULAR STRUCTURE OF AMINO ACIDS
Determination by X-Ray Diffraction Analysis

STRUKTURY AMINOKISLOT

СТРУКТУРЫ АМИНОКИСЛОТ

The Molecular Structure of Amino Acids

Determination by X-Ray Diffraction Analysis

By Galina V. Gurskaya

Institute of Crystallography
Academy of Sciences of the USSR, Moscow

Translated from Russian

CONSULTANTS BUREAU · NEW YORK · 1968

Galina Viktorovna Gurskaya was graduated from Gorki University's Department of Solid State Physics in 1958. As a graduate student, she worked on the structures of biological objects at the Institute of Crystallography of the Academy of Sciences of the USSR. The structural study of the amino acid phenylalanine was the subject of her dissertation, and she continued her work on the x-ray study of biological objects in the laboratory of B. K. Vainshtein at the Institute.

ISBN-13:978-1-4684-1568-1
DOI: 10.1007/978-1-4684-1566-7

e-ISBN-13:978-1-4684-1566-7

The original Russian text, published by Nauka Press in 1966 in Moscow for the Institute of Crystallography of the Academy of Sciences of the USSR, has been extensively revised by the author for this edition.

Галина Викторовна Гурская

Структуры аминокислот

Library of Congress Catalog Card Number 68-18821

CONTENTS

CONTENTS

PREFACE

The last decade has seen great progress in the study of the structure and functions of living organisms at the molecular level; molecular biology has become a new branch of science [1-16]. One of the major problems of molecular biology concerns the structure of proteins.

Some information about protein structure is provided by the electron microscope, by the ultracentrifuge, and by small-angle x-ray scattering; the last gives the fullest information about the atomic structure of protein molecules, but x-ray study of biological objects represents a difficult and laborious task. In this way the structures of three proteins have been established: hemoglobin, myoglobin [2, 8], and lysozyme [10]; work has begun on ribonuclease, insulin, chymotrypsin, carboxypeptidase, and so on [14-16].

A knowledge of the structures of amino acids and peptides is of considerable value in elucidating the structure and functions of proteins. Nearly all proteins consist largely of 22 principal amino acids, which are linked together in polypeptide chains. The peptide link always arises between an α-amino group and the carboxyl group in the next amino acid residue, so all polypeptide chains have the same backbone, to which are attached different radicals R (Fig. 1).

The spatial structure of a protein molecule may be said to have four levels of organization:

Primary structure, or essentially the chemical formula (the sequence and number of the amino acid residues in the polypeptide chain);

Secondary structure: organization of the chain governed by regular disposition of the residues one relative to another (for instance, parts of the chain may have the α-spiral configuration);

Tertiary structure: the general disposition of the polypeptide chain as a whole (which in some parts has a specific secondary structure);

Quaternary structure, which sometimes occurs as a mode of association of subunits having tertiary structure.

Each level has a decisive or predominant type of atomic interaction. For instance, the primary structure is entirely determined by the covalent bonds in the polypeptide main chain, while the secondary structure is maintained largely by hydrogen bonds between peptide groups. The tertiary structure is stabilized by van der Waals forces between the lateral radicals, and sometimes also by ionic bonds, disulfide bridges, and hydrogen bonds. The quaternary structure arises via local electrostatic and van der Waals interactions between groups at the surface of subunits [5, 6, 8, 11, 16].

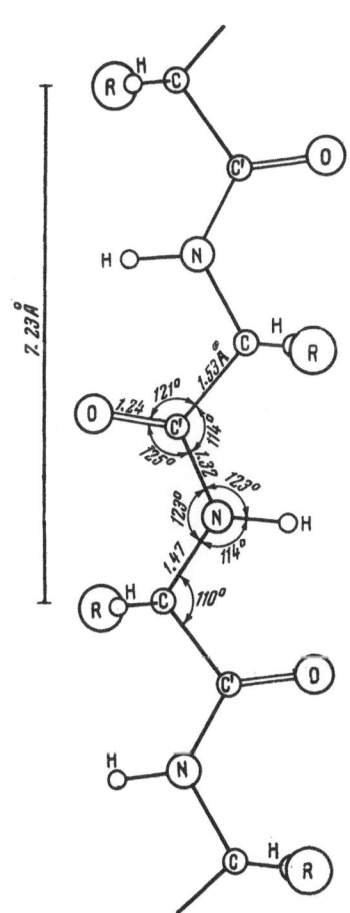

Fig. 1. General form and parameters of a polypeptide chain.

1

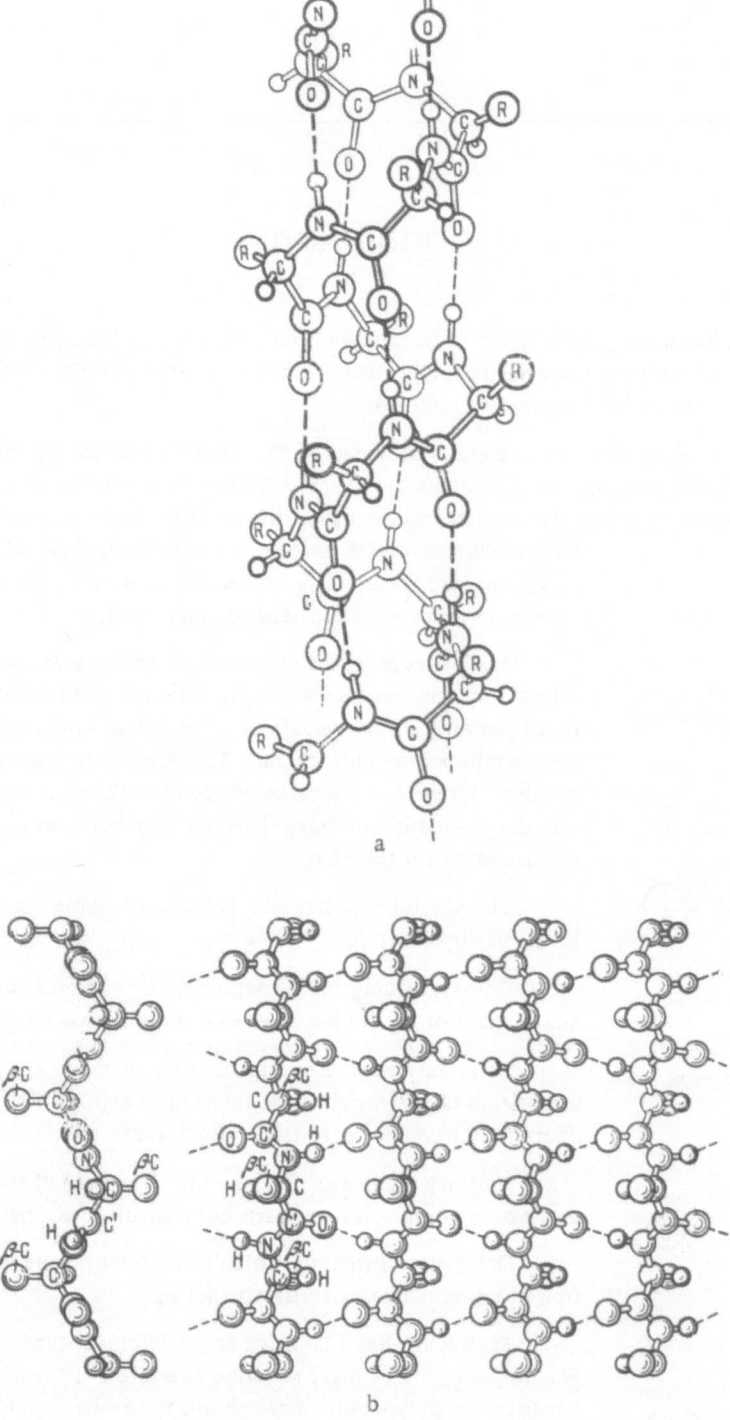

Fig. 2. Possible configurations of polypeptide chains: (a) α-spiral, as found in myoglobin and hemoglobin; (b) parallel folded layer, as in the extended β-form of keratin.

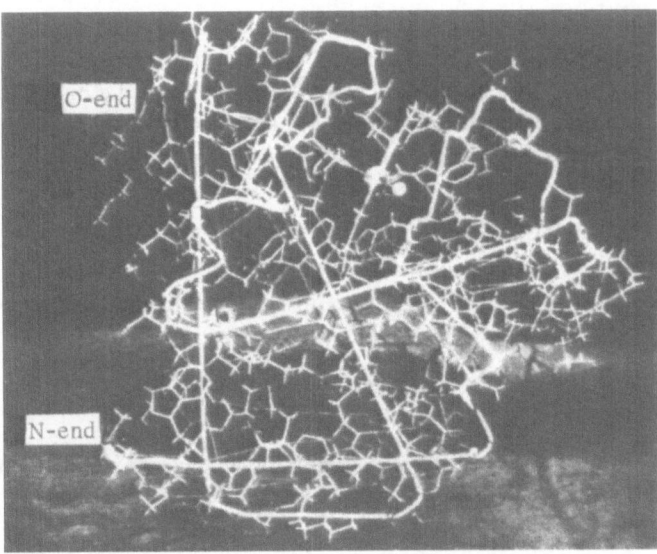

Fig. 3. Model of myoglobin molecule built up from wire
models of the amino acid residues. The heavy line indicates
the axis of the α-spiral (tertiary structure).

Amino acid structures and the chemical formula of a protein give some indication of all levels of the spatial organization. For instance, conclusions have been drawn from the structures of amino acids and peptides about the structure of the polypeptide backbone and about the size of the amide groups (Fig. 1), which in turn has provided a means of theoretical consideration of the formation of hydrogen bonds between peptide groups, which has led to the prediction of secondary structures (spiral and β-stretched configurations) [17]. Some of these have been shown to occur in globular and fibrous proteins (Fig. 2).

It was long ago supposed that the tertiary structure arises because the nonpolar (hydrophobic) groups tend to lie within the protein globule, while the polar (hydrophilic) groups tend to lie on the outside. This tendency is found to occur in myoglobin and hemoglobin. Water molecules are attached to the polar groups and form a double electrostatic layer, which screens the polar groups and reduces the free energy. Water molecules become ordered in contact with hydrophobic groups, and this reduces the entropy, so the hydrophobic groups are best placed within the structure away from contact with water, which minimizes the free energy and maximizes the entropy of the protein—water system. The van der Waals interaction is sufficiently strong to explain the compact array of nonpolar radicals within the structure, with additional stabilization by some ionic and hydrogen bonds [16].

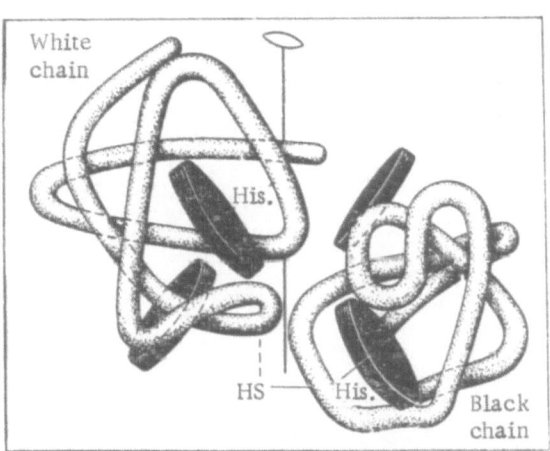

Fig. 4. Disposition of the polypeptide chains in the hemoglobin molecule (quaternary structure). The black discs indicate the positions of the four haem groups. Only two of the four polypeptide chains are shown for clarity, the others being generated by a twofold axis.

Bends in the polypeptide chains may arise from the specific molecular structures of some amino acids, e.g., proline and hydroxyproline, whose amine nitrogen forms part of the pyrrolidine ring, and which do not take up the α-spiral configuration without considerable steric hindrance. Occurrence of these usually means deviation from a regular spiral and change in the direction of the chain.

It has been found from synthetic polypeptides consisting of one type of amino acid that α-spirals are also not produced by polyserine, polythreonine, polyvaline, polyisoleucine, and polyhistidine; these residues are termed antispiral ones. However, it has been found from the statistics of the disposition of the lateral radicals in myoglobin and hemoglobin that certain kinks consist almost entirely of these residues, but that the latter nevertheless occur frequently in the α-spiral parts. A kink is found [16] always to contain histidine, glutamic acid, and aspartic acid.

This is illustrated by the structures shown in Figs. 3 and 4. The polypeptide chain of myoglobin consists of 153 residues, 70% of which are included in the α-spiral. The molecule generally is coiled in a very complicated way into a compact form (Fig. 3); the chain loses its spiral conformation near points of bending in the α-spiral.

The hemoglobin molecule is built up from four subunits (quaternary structure), each subunit being very similar to the myoglobin molecule (Fig. 4).

The significance of amino acid structures is not simply that these acids are constituents of proteins, for they occur frequently in the free state and are of prime biochemical importance [18].

Crystal structures are now known for all the principal amino acids except tryptophan. I have surveyed all the available evidence, since a great deal of interesting evidence has accumulated since Hahn's survey of 1957 [22].

This book includes material appearing up to the end of 1965, as well as some unpublished work kindly made available by workers at Madras University.

The divisions are as follows. First I consider briefly the main physiochemical properties and classification of the amino acids. Then the first and second chapters deal with the description of the crystal structures of the principal amino acids and certain others. The third chapter presents an analysis of the general trends in the structure and crystal packing.

I am indebted to B. K. Vainshtein, who proposed this work and who constantly provided advice during its compilation.

I am also indebted to G. N. Tishchenko, N. S. Andreeva, and Yu. T. Struchkov for valuable comments on the manuscript.

PROPERTIES AND CLASSIFICATION OF AMINO ACIDS

The first naturally occurring amino acids were discovered early in the 19th century; over 80 are now known, of which 22 are the principal ones in proteins, with about ten others occurring less often. The other amino acids occur in other physiologically active compounds (antibiotics, hormones, peptides, etc.) or occur in the free state in plants or animals.

All the principal amino acids had been discovered by 1936 (the last, threonine, was discovered in 1935), whereas the discovery of most of the other naturally occurring ones postdates that time (Fig. 5).

The principal ones have therefore been studied in more detail, and so we shall consider these, although many of the statements apply also to the others. Table 1 lists the principal ones with their structural formulas.

The amino acids may be divided into essential and inessential, the latter consisting of those that are not obliged to be present in the food in order to insure normal development, since they can be synthesized from other amino acids or other compounds if the supply is inadequate. The amino acids that cannot be synthesized in the body are termed essential. There is no sharp division between the two classes, since the requirement for a given amino acid is dependent on the species and also on the physiological state of the animal. However, certain ones (valine, isoleucine, leucine, lysine, methionine, threonine, tryptophan, and phenylalanine) are essential for nearly all species of animal (Table 2), and these are widely used in medicine [18].

The biologically important amino acids are amphoteric compounds, since they contain NH_2 and COOH groups attached to the α-carbon.* Hence any amino acid may be represented by the general formula

$$R-\alpha\text{-}\overset{\displaystyle NH_2}{\underset{\displaystyle H}{C}}-COOH. \tag{1}$$

The specific features are governed by the structure of R; Table 1 gives a classification in terms of R.

A different type of structural formula occurs in the imino acids proline and hydroxyproline, in which the nitrogen atom forms part of a pyrrolidine ring; but these may also be considered as α-amino acids.

The ionic form is dependent on the pH: at acid pH the ion is $NH_3{+}CH(R)COOH$, while at alkaline pH it is $NH_2CH(R)COO^-$, and at neutral pH we get the zwitterion $NH_3{+}CH(R)COO^-$. The last is also the form that usually occurs in the crystal.

Fig. 5. Discovery of naturally occurring amino acids: (1) principal, (2) others.

---

*Biologically active substances sometimes also contain β- and γ-amino acids as well, which have the NH_2 attached to the β- or γ-carbon atoms (Fig. 6).

TABLE 1. Classification of the Principal Amino Acids by Type of Radical R

I. Aliphatic Amino Acids

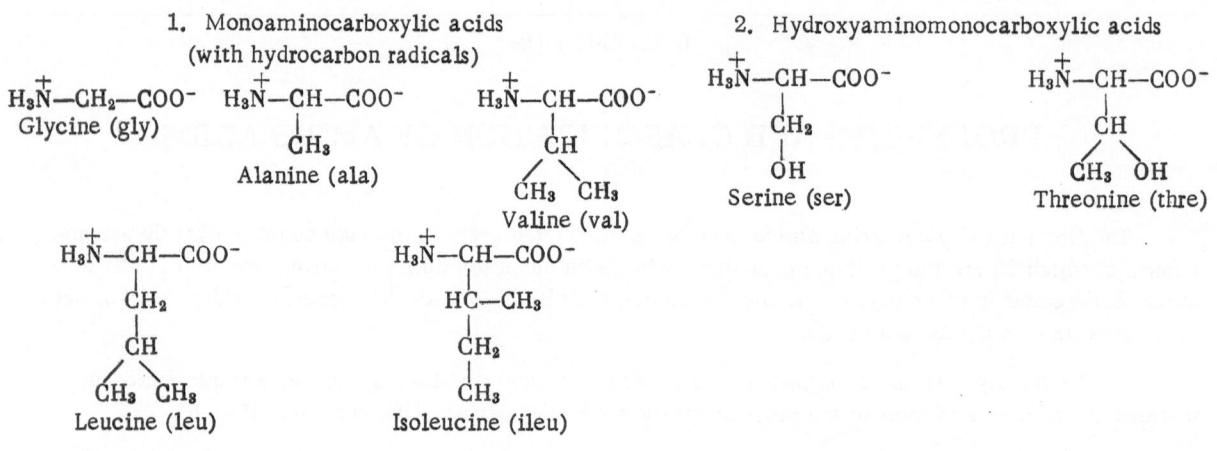

1. Monoaminocarboxylic acids
(with hydrocarbon radicals)

$H_3\overset{+}{N}$—CH_2—COO^-
Glycine (gly)

$H_3\overset{+}{N}$—CH—COO^-
|
CH_3
Alanine (ala)

$H_3\overset{+}{N}$—CH—COO^-
|
CH
/ \
CH_3 CH_3
Valine (val)

$H_3\overset{+}{N}$—CH—COO^-
|
CH_2
|
CH
/ \
CH_3 CH_3
Leucine (leu)

$H_3\overset{+}{N}$—CH—COO^-
|
HC—CH_3
|
CH_2
|
CH_3
Isoleucine (ileu)

2. Hydroxyaminomonocarboxylic acids

$H_3\overset{+}{N}$—CH—COO^-
|
CH_2
|
OH
Serine (ser)

$H_3\overset{+}{N}$—CH—COO^-
|
CH
/ \
CH_3 OH
Threonine (thre)

3. Monoaminodicarboxylic acids

$H_3\overset{+}{N}$—CH—COO^-
|
CH_2
|
COO^-
Aspartic acid (asp)

$H_3\overset{+}{N}$—CH—COO^-
|
CH_2
|
CH_2
|
COO^-
Glutamic acid (glu)

4. Amides of monoaminodicarboxylic acids

$H_3\overset{+}{N}$—CH—COO^-
|
CH_2
|
C
/ \
O NH_2
Asparagin (asp-N)

$H_3\overset{+}{N}$—CH—COO^-
|
CH_2
|
CH_2
|
C
/ \
O NH_2
Glutamine (glu-N)

5. Diaminomonocarboxylic acids

$H_3\overset{+}{N}$—CH—COO^-
|
CH_2
|
CH_2
|
CH_2
|
CH_2
|
$\overset{+}{N}H_3$
Lysine (lys)

$H_3\overset{+}{N}$—CH—COO^-
|
CH_2
|
CH_2
|
CH_2
|
NH
|
C
/ \
H_2N $\overset{+}{N}H_2$
Arginine (arg)

$H_3\overset{+}{N}$—CH—COO^-
|
CH_2
|
SH
Cysteine (cys-SH)

6. Sulfur amino acids

$H_3\overset{+}{N}$—CN—COO^-
|
CH_2
|
S
|
S
|
CH_2
|
$H_3\overset{+}{N}$—CH—COO^-
Cystine (cys)

$H_3\overset{+}{N}$—CH—COO^-
|
CH_2
|
CH_2
|
S
|
CH_3
Methionine (meth)

TABLE 1 (continued)

II. Aromatic and Heterocyclic Amino Acids

1. Aromatic

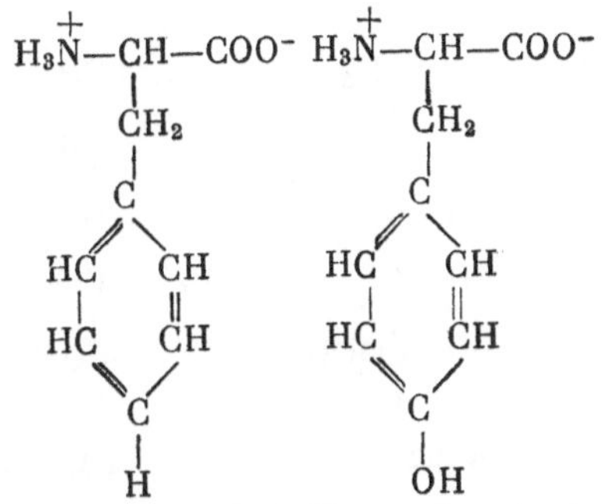

2. Heterocyclic

Phenylalanine (phal) Tryptophan (try) Tyrosine (tyr) Histidine (his)

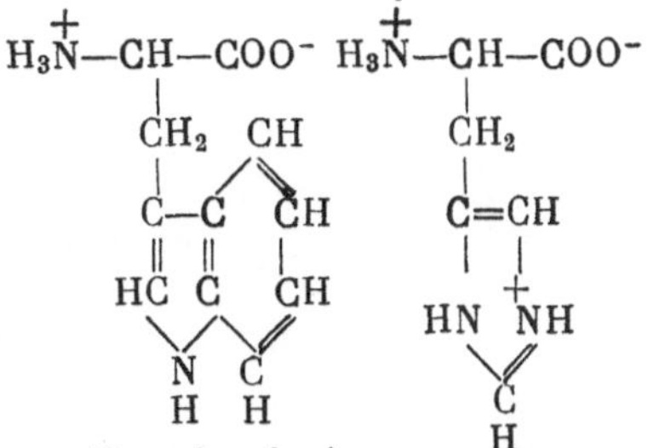

3. Imino acids (containing pyrrolidine rings)

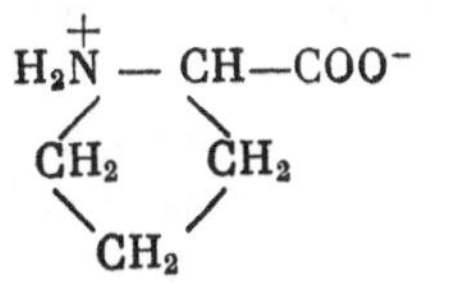

Proline (pro)

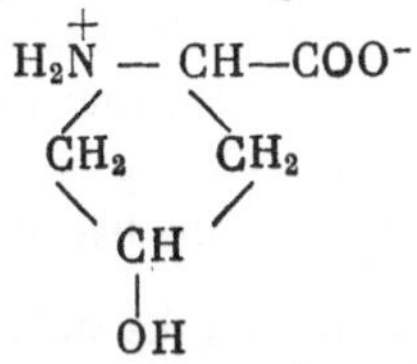

Hydroxyproline (hypro)

TABLE 2. Requirements for Amino Acids in Animal Species* [18]

Amino acid	Rat	Dog	Mouse	Chicken	Man	Tricho- monas foeus	Tetra- hymena geleii	Triboli- um con- fusum	Aedes aegypti
alanine	−	−	−	−	−	−	−	−	−
arginine	+	−	−	+	−	+	+	+	+
aspartic acid	−	−	−	−	−	−	−	−	−
valine	+	+	+	+	+	+	+	+	?
histidine	+	+	+	+	−	+	+	+	+
glycine	−	−	−	(+)	−	+	−	−	+
glutamic acid	−	−	−	(+)	−	−	−	−	−
isoleucine	+	+	+	+	+	+	+	+	+
leucine	+	+	+	+	+	+	+	+	+
lysine	+	+	+	+	+	+	+	+	+
methionine	+	+	+	+	+	+	+	+	+
hydroxyproline	−	−	−	−	−	−	−	−	−
proline	−	−	−	(+)	−	+	−	−	−
serine	−	−	−	−	−	+	+†	−	−
tyrosine	−	−	−	(+)	−	−	−	−	+‡
threonine	+	+	+	+	+	+	+	+	+
tryptophan	+	+	+	+	+	+	+	+	+
phenylalanine	+	+	+	+	+	+	+	+	+‡
cystine	−	−	−	(+)	−	−	−	−	−

*+ = essential, - = inessential, (+) = essential under certain conditions. In all animals, except man, the criterion is growth; in man the criterion is nitrogen balance.

†Serine requirements dependent on the strain.

‡Tyrosine and phenylalanine must be present together.

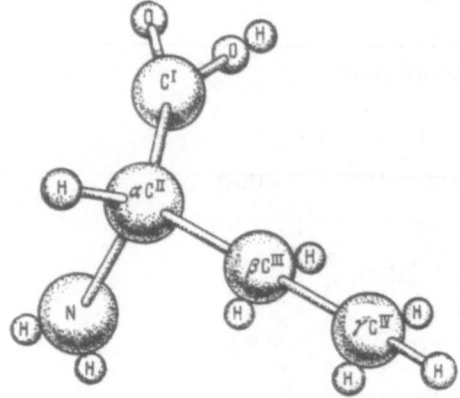

Fig. 6. Spatial disposition of the atoms
in L amino acids.

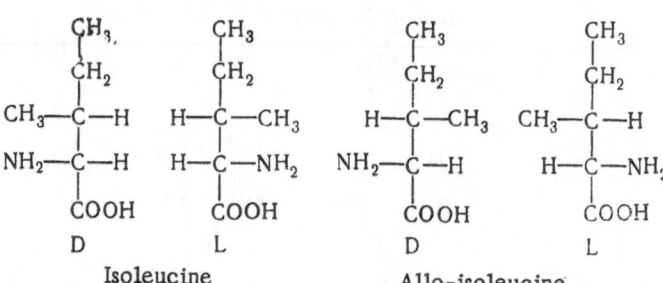

Isoleucine   Allo-isoleucine

Fig. 7. Stereoisomers of isoleucine: [D(I), L(II), D(III), L(IV)].

These polar compounds are readily soluble in water but are almost insoluble in alcohol and other organic solvents. The solubility in water is very much dependent on the length of the hydrocarbon chain; for instance, 100 g of water at 25°C will dissolve 25.3 g of glycine, 16.6 g of L-alanine, 9.1 g of L-valine, 4.1 g of L-isoleucine, and 2.4 g of L-leucine. Proline has the highest solubility in water and alcohol.

Other characteristic features are the stability at room temperature and the high melting and decomposition points [18-21].

Formula (1) shows that in every case except glycine (where R = H), the α-carbon has four different substituents and so is asymmetric; the amino acids thus exist in optically active L and D forms. Figure 6 shows the spatial disposition in the L form. The stereoisomers are completely separated in nature; all proteins consist of amino acids with the L configuration. The D forms occur mainly in the sheaths of certain bacteria and in antibiotics, i.e., poisons synthesized by cells to combat other microorganisms. The causes and means of this separation at present remain debatable.

The L and D series differ in taste, the L ones being either bitterish or tasteless, while the D ones are sweetish [18].

Certain amino acids (isoleucine, threonine, hydroxyproline) have asymmetric centers in addition to the α-carbon, but only one steric configuration is found in proteins. For example, there are the four isomers of Fig. 7 for isoleucine, but only form II occurs in proteins.

The molecular structure must be known, as well as the molecular interactions indicated by the crystalline packing, in order to explain the properties of amino acids and to reveal features common to all. X-ray methods give the fullest information on this.

The first x-ray studies of amino acids were made in 1931, when Bernal established the space groups and cell dimensions for glycine, DL alanine, D phenylalanine, L cystine, etc. [23].

The first complete structure study for the α-modification of glycine was reported in 1939 [24]. Since then, structure studies have been performed for most of the principal amino acids and for certain others; the absolute configuration of the molecule has been determined in certain cases. The following sections give brief descriptions of all amino acid structures that have so far been examined.

CHAPTER I

CRYSTAL STRUCTURES OF ALIPHATIC AMINO ACIDS

1. MONOAMINOMONOCARBOXYLIC ACIDS

Glycine (Aminoacetic Acid)

H_2NCH_2COOH

This was the first amino acid isolated from a protein (Braconno, 1820); its sweet taste led to its being given the name glycocoll, and subsequently glycine, which means the sugar of gelatin. Only afterwards was it found that glycine contains nitrogen and is an amino acid.

Glycine occurs in many proteins, and it is present in especially large amounts in silk fibroin and in collagen [18-20].

It differs from the other amino acids in not having an asymmetric carbon atom, so its solutions are not optically active; but the molecule is not planar in the crystal and can exist in two mirror-image forms.

Three crystalline modifications are known: α-, β-, and γ-glycine [23-36].

The first data on the space group and cell parameters of α-glycine were published in 1931 [23]; then, and again in 1936 [26, 27], attempts were made to deduce the atomic structure of α-glycine. A complete structure study was performed in 1939 [24, 28].

The crystals were grown from a saturated aqueous solution by slow evaporation of the solvent. They have space group $P2_1/n$ and the following unit-cell parameters: $a = 5.10$ Å ($5.1020*$), $b = 11.96$ Å (11.9709), $c = 5.45$Å (5.4575), $\beta = 111°38'$ ($111°42.3'$), $Z = 4$.

A structure model was deduced from the three-dimensional Patterson function with extensive use of Harker sections; the structure was refined from the intensities of the zonal reflections. After refinement, almost all the bond lengths and valence angles were close to the standard values, apart from the $C-N$ bond (1.39 Å), which was 0.08 Å shorter than the standard for a single $C-N$ bond.

Marsh [30] performed a more detailed structure refinement in 1957. He used the intensities of 1867 reflections produced by Mo $K\alpha$ radiation, the initial coordinates being those found by Albrecht and Corey. The initial corrections were deduced from ordinary and differential projections of the electron density; then several three-dimensional difference syntheses were used to deduce the coordinates of the hydrogen atoms and to refine the positions of the other atoms, as well as the anisotropic temperature factors. The process closed with two cycles of refinement by least squares. The result for $R(hkl)$ was 6.3%. The interatomic $C-C$, $C-N$, and $C-O$ distances were determined to ~ 0.005 Å; the $C(N)-H$ distances were determined to ~ 0.06 Å, and the valence angles to $\sim 0.3°$. Table 3 gives the parameters of the atoms.

Figure 8 (see also Table 4) shows the structure of the molecule in α-glycine

The bond lengths and valence angles agree well with those found for other amino acids.

---

*The figures in parentheses are those found by Marsh [29, 30].

TABLE 3. Parameters of the Atoms in the Structure of α-Glycine [Temperature Factors of the Form $T_j = \exp-(B_{11}h^2 + B_{22}k^2 + B_{33}l^2 + B_{12}hk + B_{13}hl + B_{23}kl)$]

Atom	x	y	z	$B_{11} \cdot 10^4$	$B_{22} \cdot 10^4$	$B_{33} \cdot 10^4$	$B_{12} \cdot 10^4$	$B_{13} \cdot 10^4$	$B_{23} \cdot 10^4$
C_1	0.07542	0.12478	0.06605	157	22	115	-10	150	-10
C_2	0.06536	0.14499	0.78711	156	32	118	20	130	16
N	0.30135	0.08980	0.74113	164	38	108	34	130	12
O_1	0.30583	0.09427	0.23553	205	50	115	45	138	28
O_2	0.85224	0.14154	0.10711	185	56	190	8	230	-20
H_1	0.286	0.100	0.570						
H_2	0.457	0.119	0.837						
H_3	0.298	0.020	0.763						
H_4	0.080	0.220	0.771						
H_5	0.898	0.118	0.671						

A distinctive feature is that the molecule is not planar, the N atom deviating from the plane of the COOH group and C_2 by 0.436 Å. In spite of the molecular asymmetry, the crystal as a whole is inactive, because the packing is centrosymmetric. The distribution of the hydrogen atoms corresponds to the form $^+H_3NCH_2COO^-$ (Fig. 9).

The dipolar molecules are linked via N−H...O hydrogen bonds into double layers parallel to the ac plane; each single layer has strong hydrogen bonds of lengths 2.768 and 2.850 Å (Fig. 10a), while weak bifurcated hydrogen bonds (2.949 and 3.074 Å) join the layers into pairs (Figs. 10b and 11). The double layers are held together by van der Waals forces, which explains the good cleavage on (010).

The addition of ethanol to a saturated aqueous solution produces the second form, but it can be kept for a time in a dry atmosphere.

Fischer described β-glycine in 1905; in 1931 Bernal [23] determined the space group and cell parameters, but the space group subsequently found by Ksanda [31] and Iitaka [32] was different. Iitaka [33] performed a full structural study of β-glycine in 1961; the space group was found as $P2_1$, the cell parameters being $a = 5.077_4$ Å; $b = 6.267_6$ Å; $c = 5.379_9$ Å; $\beta = 113°12'$; $Z = 2$. The entire analysis was based on the intensities of the h0l and 0kl refractions as measured with a diffractometer.

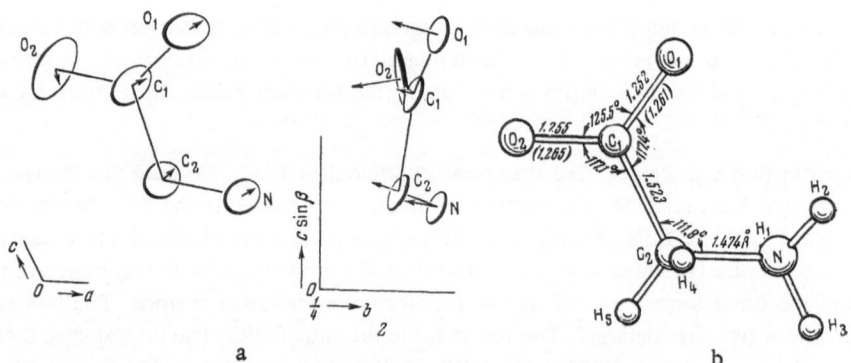

Fig. 8. Structure of the glycine molecule in α-glycine: (a) ellipsoids of thermal vibration (1) seen along b axis, (2) seen along c axis, the arrows being perpendicular to the plane of the molecule. Apart from C_1, these coincide with the principal axes of the ellipsoids. The lengths of the arrows are proportional to the axes as projected on the plane of the molecule; (b) bond lengths and valence angles. The quantities in parentheses are the values corrected for the thermal motion of the atoms.

TABLE 4. Lengths of Intramolecular Bonds and Valence Angles in the Structure of α-Glycine

$N-H_1$	0.92Å	C_2-N-H_1	110°	$N-C_2-H_4$	109°
$N-H_2$	0.85	C_2-N-H_2	110	$N-C_2-H_5$	110
$N-H_3$	0.85	C_2-N-H_3	111	$H_4-C_2-H_5$	112
				H_1-N-H_2	106
C_2-H_4	0.91	$C_1-C_2-H_4$	106	H_1-N-H_3	106
C_2-H_5	0.91	$C_1-C_2-H_5$	108	H_2-N-H_3	113

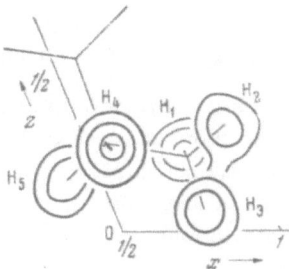

Fig. 9. Projection of sections of the difference synthesis constructed to determine the positions of the hydrogen atoms in α-glycine. The lines are at intervals of 0.25 e/Å³, the first being at 0.25 e/Å³.

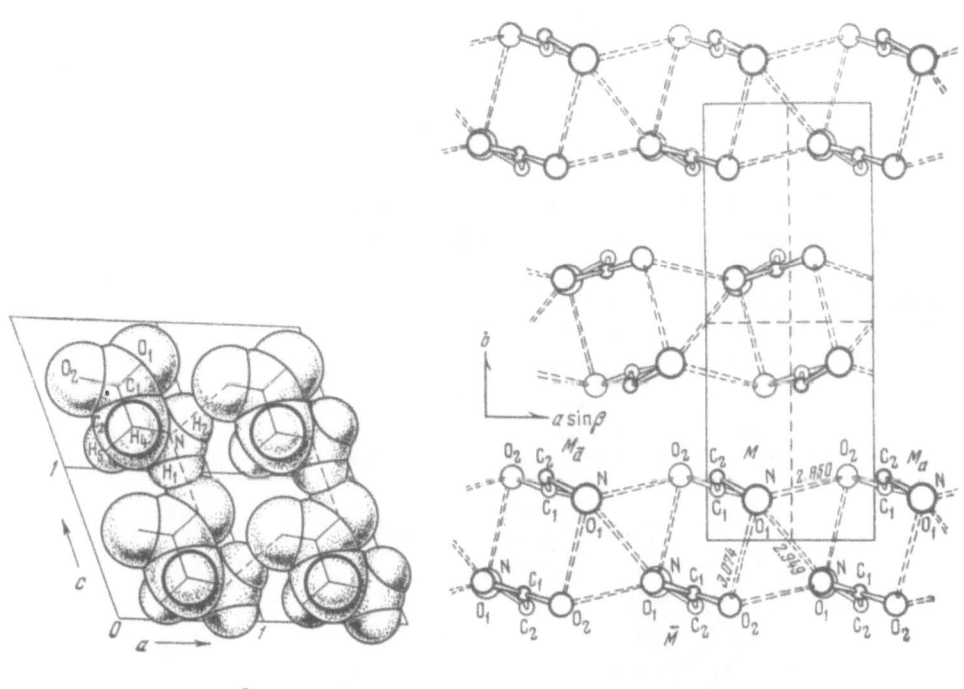

Fig. 10. Structure of α-glycine: (a) mode of packing of molecules in single layers; (b) view of structure along c axis.

The model was chosen on the basis of close similarity of the α and β forms as regards constants a and c, with b roughly halved in β-glycine. The structure was refined from differential electron-density projections. The coordinates of the hydrogen atoms were deduced from chemical crystallography. Allowance for the anisotropic thermal motion gave an R of 4.6% for the h0l reflections and 4.3% for the 0kl ones. The interatomic distances were determined to $\sim$0.015 Å and the valence angles to $\sim$0.3°. Table 5 gives the atomic parameters.

The molecular structure in the β form is similar to that of the α form, but the nitrogen atom is here 0.583 Å from the plane of the rest of the molecule (Fig. 12).

The packing in the crystal is also somewhat similar. Strong N—H...O hydrogen bonds (2.758 and 2.833 Å) join the molecules into single layers parallel to (010) (Fig. 13a), while the single layers are linked by bifurcate hydrogen bonds (3.002 and 3.022 Å), not into pairs (as in α-glycine) but into a three-dimensional framework (Fig. 13b).

The third form (γ-glycine) is stable at room temperature and goes over to α-glycine only on heating to 165°C. This was first made by Iitaka in 1954 [34] by slow cooling of an aqueous solution containing acetic acid. The space group is $P3_1$ or $P3_2$, the cell parameters being a = 7.037 Å; c = 5.483 Å; Z = 3.

The trial model [35, 36] was chosen on the basis of equality of the c lattice constants of the α and γ forms; a Harker section (z = 1/3) of the Patterson function was also used. The model was refined via two-dimensional electron-density projections, differential projections, and finally several three-dimensional least-squares cycles. The coordinates of the hydrogen atoms were deduced from arguments from chemical crystallography. The final R(hk$l$) was 10.8% for the entire three-dimensional set of intensities. The interatomic distances were determined to $\sim$0.011 Å and the valence angles to $\sim$0.7°. Table 6 gives the parameters of the atoms.

TABLE 5. Parameters of the Atoms in the Structure of β-Glycine (Temperature Factors of the Form $T_j = \exp - Bj \sin^2 \theta / \lambda^2$, $Bj = B_1 + B_2 \sin^2 (\omega - \varphi)$, in Which ω Are the Polar Coordinates of the Points of the Reciprocal Lattice Relative to the c Axis)

Atom	x	y	z	B_j(Å^2) for h0l projection	Bj(Å^2) for 0kl projection
N	0.3522	-0.0440	-0.2619	$1.96 + 0.40 \sin^2 (\omega + 70°)$	$2.32 + 1.73 \sin^2 \omega$
O_1	0.3772	0.0270	0.2420	$2.23 + 0.30 \sin^2 (\omega - 12°)$	$1.97 + 3.15 \sin^2$
					$(\omega + 17°)$
O_2	-0.0896	0.0773	0.0970	$1.69 + 1.89 \sin^2 (\omega + 70°)$	3.36
C_1	0.1378	0.0532	0.0633	$2.63 + 0.10 \sin^2 \omega$	2.63
C_2	0.1145	0.0719	-0.2265	2.83	$2.83 + 0.96 \sin^2 \omega$
$H_1(N)$	0.540	0.012	-0.135	2	3.44
$H_2(N)$	0.340	-0.031	-0.449	2	3.44
$H_3(N)$	0.337	-0.203	-0.216	2	3.44
$H_1(C_2)$	0.125	0.241	-0.274	2	3.44
$H_2(C_2)$	-0.091	0.008	-0.362	2	3.44

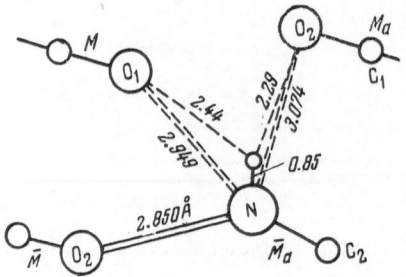

Fig. 11. Formation of bifurcate hydrogen bonds in α-glycine.

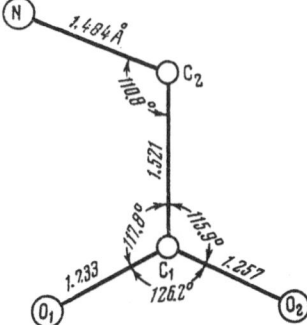

Fig. 12. Structure of the
molecule in β-glycine.

The γ form has very much the molecular configuration of the other forms (Table 7) except that the C_1-O_2 bond is somewhat shortened and the N atom deviates from the plane of the molecule by only 0.309 Å.

The mode of packing is indicated by Fig. 14. Hydrogen bonds (2.801 and 2.817 Å) join the molecules into spirals around the threefold axes parallel to the c axis. The spirals are linked by 2.970 Å hydrogen bonds and electrostatic forces between adjacent NH_3^+ and COO^- groups.

The difference between the three forms in the orientation of the C—N bonds relative to the plane of the COOH group, the angles of the C—N bonds relative to that plane being 18.6°, 24.8°, and 12.8° for the α, β, and γ forms, respectively. The stability series is α, γ, β; conversion of the β and γ forms to the α form is unexpected, because the molecule has to alter in shape. A crystal of α-glycine has a center of symmetry, since it contains two mirror-image forms of the molecule, whereas the β and γ forms contain only one of the possible forms, so their crystals lack a center of symmetry. No α→β and α→γ transitions have been observed.

TABLE 6. Parameters of the Atoms in the Structure of
γ-Glycine (Temperature Factors of the Form
$T_j = \exp - B_j \sin^2 \theta / \lambda^2$)

Atom	x	y	z	B_j, $Å^2$
N	0,2414	0.0263	0.5035	2.03
O_1	0.2325	0.0083	0.0139	2.49
O_2	0.5425	0.0011	-0.0150	3.08
C_1	0.3929	0.0012	0.1033	2.00
C_2	0.4010	-0.0222	0.3794	1.98
$H_1(N)$	0.248	0.013	0.686	1.50
$H_2(N)$	0.274	0.183	0.458	1.50
$H_3(N)$	0.084	-0.079	0.441	1.50
$H_1(C_2)$	0.567	0.089	0.446	1.50
$H_2(C_2)$	0.369	-0.186	0.427	1.50

TABLE 7. Intramolecular Distances and Valence
Angles for γ-Glycine

$N-C_2$	1.491 Å	C_1-C_2-N	110.8°
C_2-C_1	1.527	$O_1-C_1-O_2$	125.4
C_1-O_1	1.254	$C_2-C_1-O_1$	118.3
C_1-O_2	1.237	$C_2-C_1-O_2$	116.2

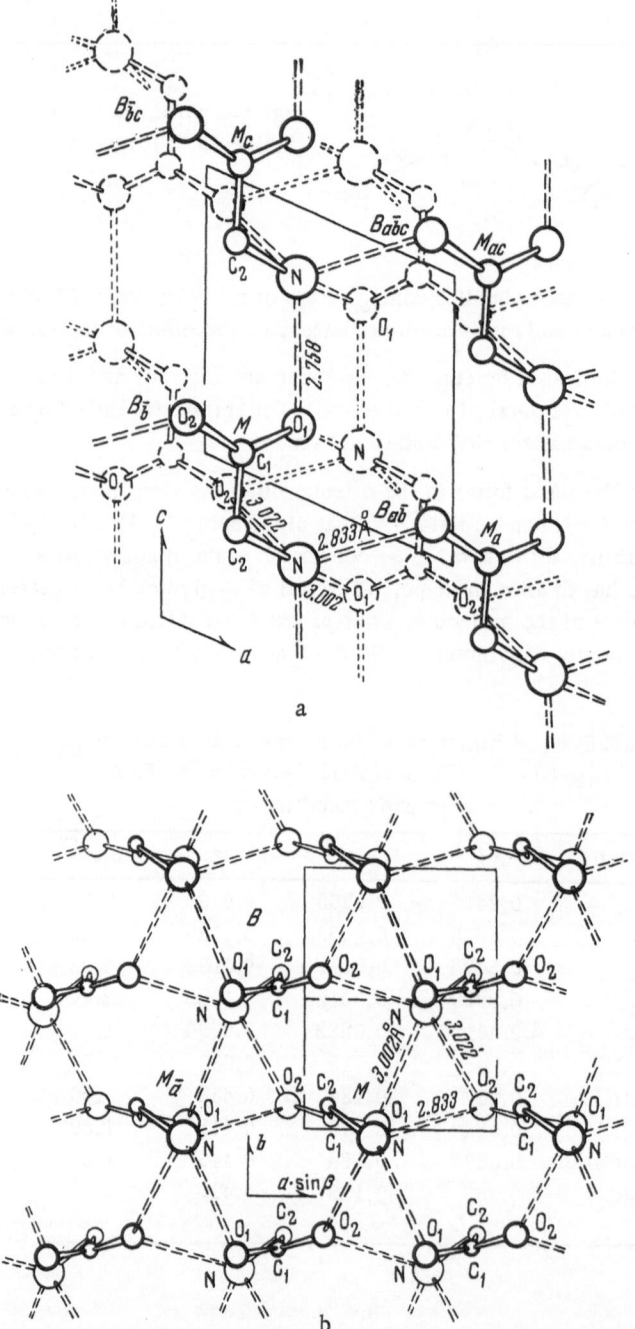

Fig. 13. Schematic representation of the structure of β-glycine: (a) view along b axis, (b) view along c axis.

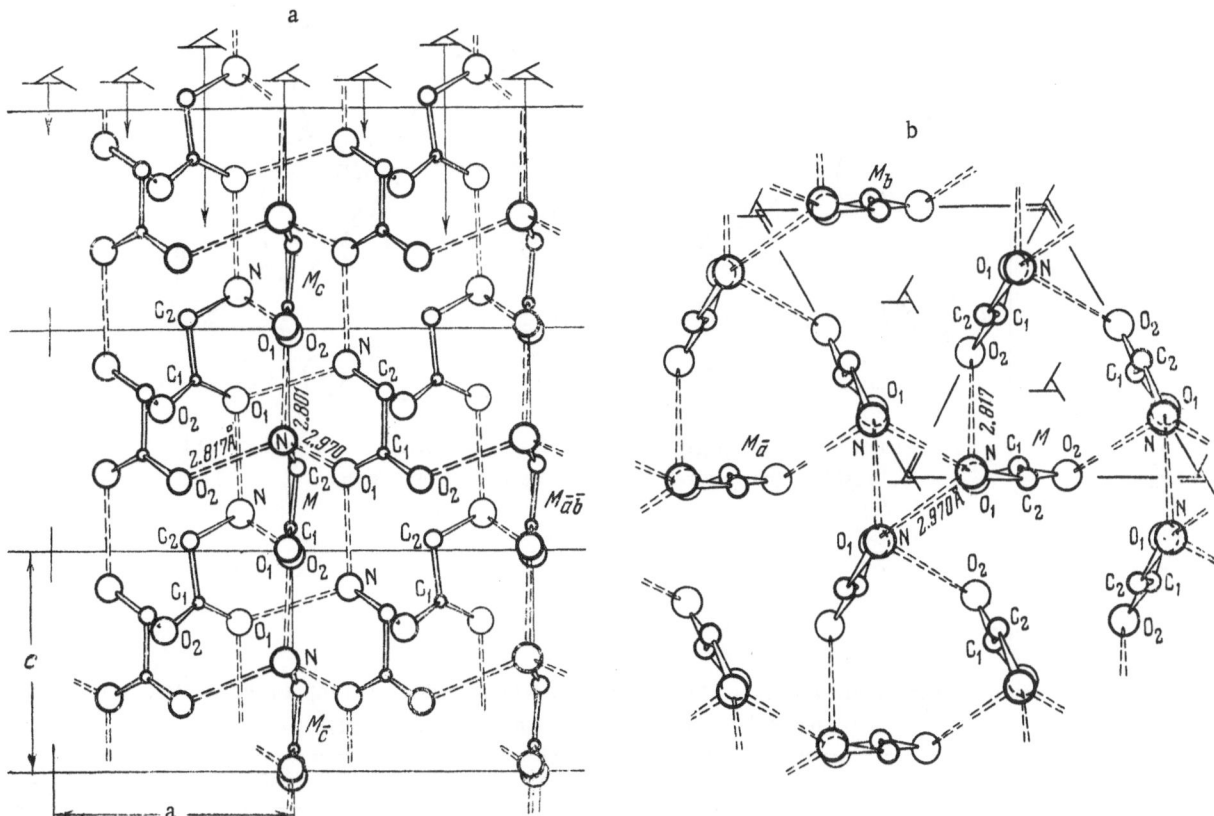

Fig. 14. Schematic representation of the structure of γ-glycine: (a) view along a axis, (b) view along c axis.

TABLE 8. Coordinates of the Main Atoms
in Ni(gly)$_2$ · 2H$_2$O

Atom	x	y	z
Ni	0.000	0,000	0.000
O$_1$	0.074	0,272	0.883
O$_2$	0.308	0 505	0.790
O$_3$	0.132	0.859	0.779
N	0.276	0.025	0.996
C$_1$	0.232	0.345	0.871
C$_2$	0.322	0.233	0.959

The structure has also been examined for certain salts. In 1945 Stosick described the structure of nickel glycinate dihydrate [37], the crystals being grown from aqueous solution by slow evaporation at a constant temperature. The space group is P2$_1$/c, the cell parameters being a = 7.60 Å; b = 6.60 Å; c = 9.63 Å; β = 63°25'; ρ_{meas} = 1.86 g/cm^3; Z = 2[Ni(gly)$_2$ · 2H$_2$O].

The model was drawn up by the heavy-atom method from Patterson projections; no detailed refinement was performed, so the accuracy is low (Table 8).

Figure 15a shows the interatomic distances and valence angles for the glycine part; Fig. 15b gives a general view of the structure, which consists of distorted octahedral groups, each consisting of a nickel atom sur-

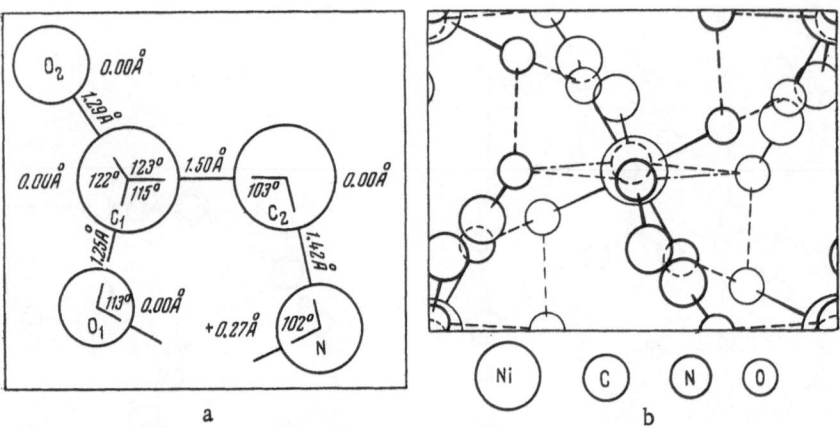

Fig. 15. Structure of nickel glycinate dihydrate: (a) structure of amino acid
residue; (b) general view of structure along a axis.

TABLE 9. Parameters of the Main Atoms in Copper Glycinate Monohydrate [Temperature Factors of the Form
$T_j = \exp - (B_{11}h^2 + B_{22}k^2 + B_{33}l^2 + 2B_{12}hk + 2B_{13}hl + B_{23}kl)]$

Atom	x	y	z	$B_{11} \cdot 10^4$	$B_{22} \cdot 10^4$	$B_{33} \cdot 10^4$	$B_{12} \cdot 10^4$	$B_{13} \cdot 10^4$	$B_{23} \cdot 10^4$
Cu	0.1031	0.3403	0.4013	58	161	20	17	1	4
O_1	0.0813	0.5772	0.2903	46	179	25	13	3	-12
O_2	0.1242	0.6286	0.1314	78	205	23	20	12	11
O_3	0.0104	0.5568	0.4919	102	178	23	40	9	11
O_4	-0.0664	0.5797	0.6444	84	341	31	20	20	0
O_5	0.2878	0.5558	0.4530	74	588	50	-88	-35	60
N_1	0.1842	0.1152	0.3025	36	159	32	15	-4	11
N_2	0.1196	0.0989	0.5176	64	266	34	37	12	28
C_1	0.1938	0.2519	0.2074	73	130	36	1	3	6
C_2	0.1285	0.5047	0.2085	22	129	35	-10	-10	2
C_3	0.0623	0.2162	0.6065	43	476	28	5	74	37
C_4	-0.0045	0.4676	0.5803	43	219	22	1	2	-1

rounded by two glycine residues and two water molecules. The nickel has four coplanar and nearly mutually
perpendicular covalent bonds to O_1 (2.08 Å) and the nitrogen atoms (2.09 Å), as well as two bonds to O_3 in the
water molecules (2.12 Å), which are perpendicular to the plane of the other four bonds. The octahedra are
linked via $N-H...O_2$ bonds (2.96 and 3.13 Å) and O_3-H-O_1 bonds (2.72 Å).

Fairly detailed studies have been made [38-41] of the copper salts of glycine, which can [39, 40] occur
in anhydrous form or as monohydrate or dihydrate.

IR spectra indicate that the dihydrate has a structure resembling that of the nickel dihydrate [39], but it
has been supposed [40] that the crystals called copper glycinate dihydrate [39] were actually ones of a second
(platy) form of the monohydrate. A complete structure study of this form has been reported [40].

TABLE 10. Interatomic Distances and Valence Angles in the Structure of Copper Glycinate Monohydrate

$Cu-O_1$	1.95_7Å	N_2-N_2	2.98_9	$Cu-N_1-C_1$	109.3
$Cu-O_3$	1.94_6	O_1-O_2	2.83_1	$Cu-N_2-C_3$	109.6
$Cu-N_1$	1.98_4	O_1-Cu-N_1	85.0°	$Cu-O_1-C_2$	115.3
$Cu-N_2$	2.02_1	O_3-Cu-N_2	85.4	$Cu-O_3-C_4$	115.8
$Cu-O_5$	2.40_4	N_1-Cu-N_2	96.6	$N_1-C_1-C_2$	112.6
N_1-C_1	1.47_3	O_1-Cu-O_3	92.9	$N_2-C_3-C_4$	111.3
N_2-C_3	1.48_4	$N_1-Cu-O_{2B0\bar{1}0}$	93.0	$C_1-C_2-O_1$	117.4
C_1-C_2	1.49_8	$N_2-Cu-O_{2B0\bar{1}0}$	87.3	$C_1-C_2-O_2$	118.3
C_3-C_4	1.54_1	$O_1-Cu-O_{2B0\bar{1}0}$	91.2	$O_1-C_2-O_2$	124.3
C_2-O_1	1.27_5	$O_3-Cu-O_{2B0\bar{1}0}$	82.5	$C_3-C_4-O_3$	117.5
C_2-O_2	1.22_6	O_5-Cu-O_2	170.9	$C_3-C_4-O_4$	119.7
C_4-O_3	1.29_1	N_1-Cu-O_5	95.9	$O_3-C_4-O_4$	112.8
C_4-O_4	1.24_3	N_2-Cu-O_5	89.6	$Cu_{B000}-O_2-C_2$	112.6
$Cu...O_{2B0\bar{1}0}$	2.74_2	O_1-Cu-O_5	91.7		
N_1-O_1	2.66_3	O_3-Cu-O_5	88.8		
N_2-O_2	2.69_1				

The crystals of the monohydrate [39, 41] have space group $P2_12_12_1$ and cell parameters as follows: $a = 10.78 (10.86)$ Å*; $b = 5.208 (5.220)$ Å; $c = 13.47 (13.50)$ Å; $Z = 4[Cu(gly)_2 \cdot H_2O]$.

A preliminary model was drawn up by the heavy-atom method; the structure was refined by least squares. The final $R(hkl)$ was 9.1%. The Cu–X(O,N) distances were determined to ~0.010 Å, while the C–X(O,N,C) were determined to ~0.016 Å. Table 9 gives the parameters of the main atoms.

Figure 16 indicates the structure, which consists of groups, each of which is a Cu^{2+} ion surrounded by an octahedron of O_1, O_3, N_1, and N_2 from two glycine residues (which have the cis configuration relative to the Cu^{2+}), O_5 from the water molecule, and O_2 from a third glycine molecule. The coordinate octahedron is elongated.

The atoms O_1, O_3, N_1, and N_2 are not strictly coplanar, the mean deviation from the best plane being ~0.028 Å. The copper ions deviate from this plane by 0.05 Å toward the water molecule.

The α-carbon and the carboxyl group are coplanar in the glycine residues, while N_1 and N_2 deviate from the planes of their corresponding carboxyl groups by 0.103 and 0.162 Å. The angle between the planes of the two residues is 5.4°.

The principal interatomic distances and valence angles are given in Table 10.

The group is formed from three glycine molecules (not the usual two), which produces the specific organization of the structure, in which the $O_{2B0\bar{1}0}...Cu$, $O_2...Cu_{B000}$ bonds join the groups into spirals parallel to the c axis. Within the spirals there are $N_{1M010}---O_1$ (3.03 Å) and $N_{2M010}---O_3$ (3.09 Å) hydrogen bonds. The spirals are linked one to another only via the hydrogen bonds O_5---O_{4A011} (2.80 Å), O_5---O_{2C010} (2.76 Å), $O_4---N_{1A\bar{1}01}$ (2.99 Å).

Lindquist and Rosenstein [42] have examined the structure of Fe(II) sulfate-glycinate pentahydrate. The crystals were grown from a solution of the salt by addition of excess sulfuric acid; oxidation was avoided by the use of an inert atmosphere. The space group is $P\bar{1}$, the lattice constants being $a = 6.86$ Å; $b = 13.6$ Å; $c = 6.07$ Å; $\alpha = 96.1°$; $\beta = 96.8°$; $\gamma = 92.5$; $Z = 2$.

---

*The values in parentheses are from [39]; the others are from [41].

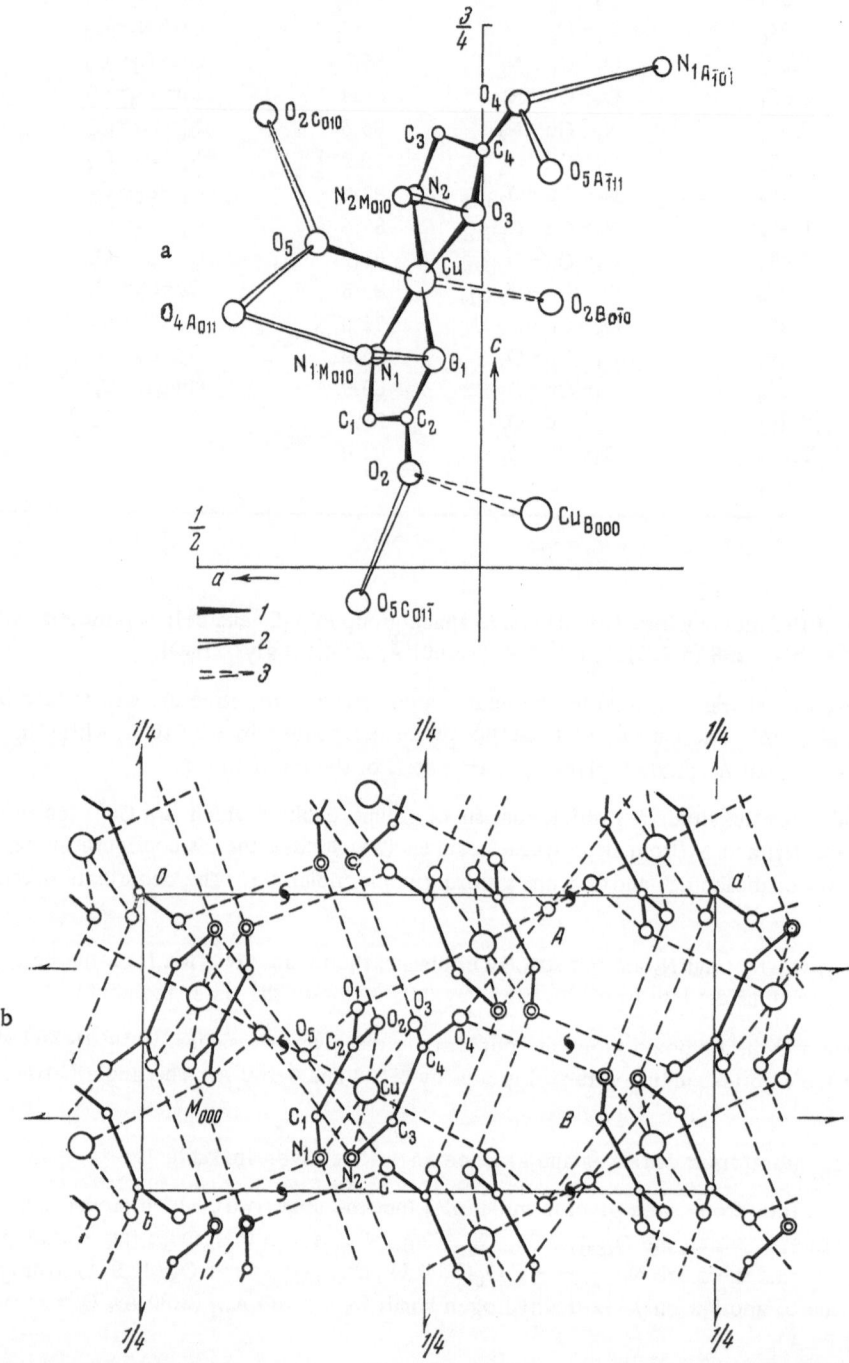

Fig. 16. Structure of copper glycinate monohydrate: (a) structure and environment of the asymmetric unit: (1) covalent bonds, (2) hydrogen bonds, (3) Cu–O interactions; (b) schematic representation of the structure as seen along the c axis.

TABLE 11. Preliminary Atomic Coordinates
in the Structure of Fe(II) Sulfate-Glycinate
Pentahydrate

Atom	x	y	z
O_1	0.178	0.055	0.708
O_2	0.104	0.133	0.205
O_3	0.759	0.085	0.958
O_4	0.488	0.123	0.287
O_5	0.770	0.140	0.560
O_6	0.655	0.285	0.403
O_7	0.460	0.205	0.670
O_8	0.169	0.383	0.535
O_9	0.633	0.463	0.681
O_{10}	0.169	0.343	0.067
O_{11}	0.663	0.472	0.869
C_1	0.023	0.322	0.858
C_2	0.149	0.398	0.737
N	0.003	0.345	0.100
S	0.588	0.1885	0.487
Fe_1	0	0	0
Fe_2	0.500	0.500	0

TABLE 12. Coordinates of the Main
Atoms in the Structure of Diglycine
Hydrobromide

Atom	x	y	z
Br^-	0.427	0.035	0.832
C_1	0.340	0.380	0.849
C_2	0.355	0.427	0.080
C_3	0.719	0.270	0.506
C_4	0.317	0.170	0.381
N_1	0.531	0.427	0.168
N_2	0.394	0.193	0.129
O_1	0.719	0.112	0.301
O_2	0.454	0.335	0.833
O_3	0.717	0.210	0.610
O_4	0.662	0.286	0.304

The heavy atoms (Fe and S) were located from a Patterson projection; the coordinates of the light atoms were deduced from electron-density projections and differential syntheses. No detailed refinement was performed. The R(hk0) and R(0kl) were 14 and 18%. Table 11 gives the preliminary coordinates.

Figure 17 illustrates the structure, which contains the complex cations $[Fe \cdot 6H_2O]^{+2}$ and $[Fe \cdot 4H_2O \cdot 2$ $(^-OOCCH_2NH_3^+)]^{+2}$; in the latter the Fe^{2+} is linked to the two glycine residues via oxygen atoms in the carboxyl groups, the residues having the trans configuration.

The accuracy was too low to allow a discussion of the bond lengths and valence angles, but it is clear that the cations are linked to the anions via a fairly complicated system of hydrogen bonds.

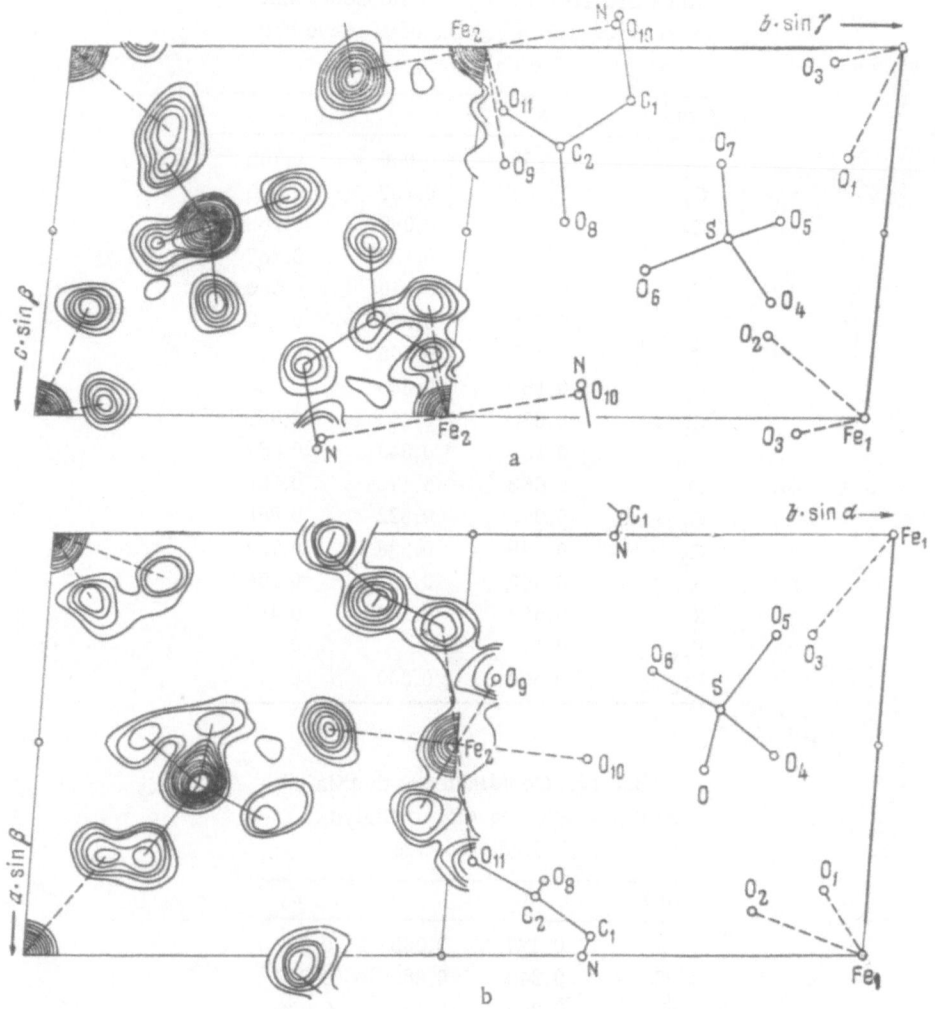

Fig. 17. Projections of the structure of Fe(II) sulfate-glycinate pentahydrate: (a) along a axis, (b) along c axis.

The structure of diglycine HBr was determined in 1956 [43]: space group $P2_12_12_1$, cell parameters a = 8.21 Å; b = 18.42 Å; c = 5.40 Å; ρ_{meas} = 1.94 g/cm^3 and Z = 4 [2gly · HBr].

A model was derived by the heavy-atom method, the coordinates (Table 12) being refined by differential electron-density projections, which gave R as 9.1% for hk0, 10.2% for 0kl, and 13.4% for h0l. The distances were found to ~0.03 Å and the angles to ~3°.

Figure 18 shows the final values of the distances and angles. The molecules take two forms: one planar within the error of measurement, the other having the nitrogen atom 0.49 Å from the plane of the rest of the molecule. All bond lengths agree well with those found for other amino acids, except C—N, which is higher.

There is a three-dimensional network of hydrogen bonds (Fig. 19); each N_1 (bent molecules) forms bonds to two bromine atoms and to O_4 in a carboxyl group, while N_2 (planar molecules) forms two ordinary hydrogen bonds to Br⁻ ions and O_4 atoms, as well as a bifurcated bond to O_2 and O_4.

There are also very strong O_1...O_3 (2.46 Å) hydrogen bonds between carboxyl groups of the two types of molecule. The structure of the carboxyl groups indicates that this bond is formed only via the hydrogen atom joined to O_1, so the planar molecules are present as $^+NH_3CH_2COO^-$, while the others are present as $^+NH_3CH_2COOH$.

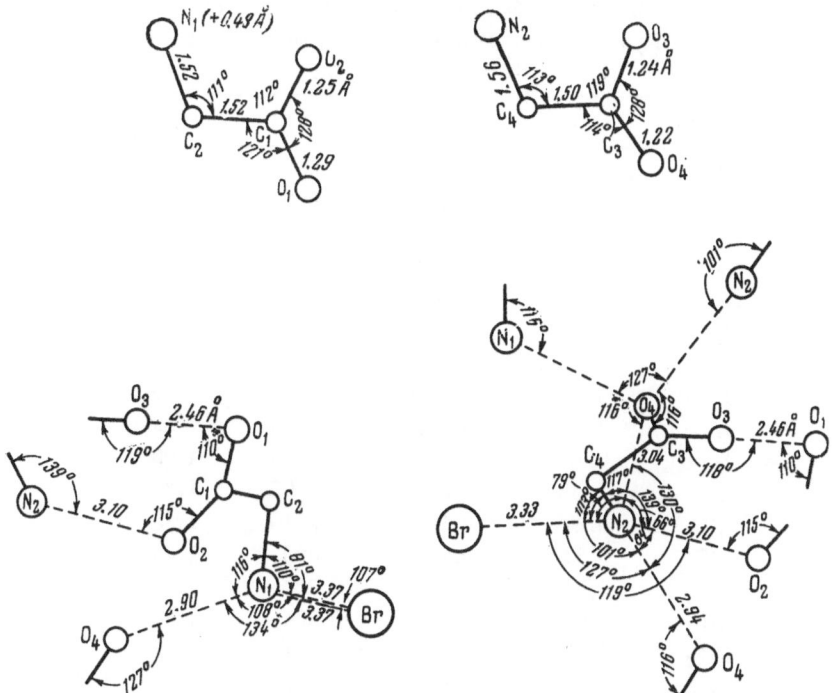

Fig. 18. Molecular form in the structure of diglycine hydrobromide.

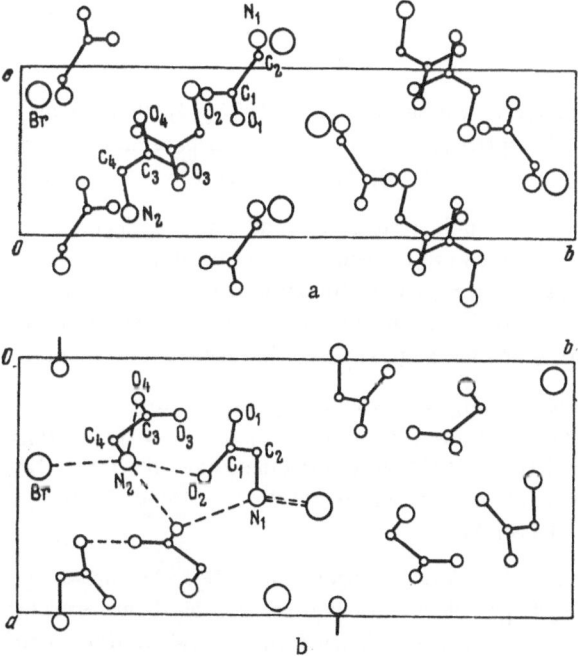

Fig. 19. Schematic representation of the structure of diglycine hydrobromide: (a) view along a axis; (b) view along c axis.

TABLE 13. Atomic Parameters in the Structure of Di-glycine Hydrochloride (Temperature Factors of the Form $T_j = \exp - B_j \sin^2\theta/\lambda^2$)

Atom	x	y	z	B_j, $\overset{\circ}{A}^2$
Cl$^-$	0.4251	0.0337	0.8373	1.79
C$_1$	0.3559	0.3819	0.8624	1.64
C$_2$	0.3626	0.4345	0.0752	1.72
C$_3$	0.7272	0.2730	0.5150	1.64
C$_4$	0.3227	0.1667	0.3640	1.67
N$_1$	0.5347	0.4349	0.1843	2.18
N$_2$	0.4005	0.1916	0.1175	1.85
O$_1$	0.7207	0.1115	0.2722	1.92
O$_2$	0.4618	0.3341	0.8260	2.29
O$_3$	0.7200	0.2105	0.6176	1.93
O$_4$	0.6623	0.2888	0.3003	1.86
H$_1$(C$_1$)	0.277	0.414	0.221	1.85
H$_2$(C$_1$)	0.332	0.491	0.009	1.85
H$_3$(N$_1$)	0.541	0.468	0.334	1.85
H$_4$(N$_1$)	0.570	0.384	0.224	1.85
H$_5$(N$_1$)	0.612	0.456	0.051	1.85
H$_6$(C$_4$)	0.236	0.122	0.320	1.85
H$_7$(C$_4$)	0.419	0.150	0.492	1.85
H$_8$(N$_2$)	0.319	0.220	0.013	1.85
H$_9$(N$_2$)	0.437	0.147	0.020	1.85
H$_{10}$(N$_2$)	0.500	0.224	0.155	1.85
H$_{11}$(O$_1$)	0.720	0.149	0.403	1.85

The structure of the hydrochloride was determined in 1957 [44], the crystals being isomorphous with those of the hydrobromide and having also space group P2$_1$2$_1$2$_1$, the cell parameters being a = 8.15 Å; b = 18.03 Å; c = 5.34 Å; ρ_{meas} = 1.581 g/cm^3, Z = 4 [2gly · HCl], ρ_X = 1.579 g/cm^3.

This was the first occasion on which a structure model was deduced via the three-dimensional minimization function. The coordinates were refined via electron-density projections, differential syntheses, and finally three-dimensional least squares. The coordinates of the hydrogen atoms were deduced from crystallographic arguments and via differential electron-density projections. The final R(hk$l$) was 10.4%. The distances were determined to ~0.01 Å and the angles to ~0.7°. Table 13 gives the parameters.

There are two types of molecule (Fig. 20), one nearly planar (nitrogen atom 0.04 Å from plane) and the other with the nitrogen atom 0.33 Å from the plane. The disposition of the hydrogen atoms indicates that the planar molecules have the form $^+$NH$_3$CH$_2$COO$^-$ while the others have the form $^+$NH$_3$CH$_2$COOH. In both cases the C−C and C−N bond lengths (1.48 and 1.52 Å, respectively) differ substantially from the standards corresponding to single bonds (1.54 and 1.47 Å).

There is a three-dimensional network of hydrogen bonds (Fig. 21), the strongest (2.57 Å) being the O−H...O$_3$ ones arising from the hydrogen of the carboxyl groups. N$_1$ (bent form) has a bond to O$_4$ (2.90 Å) and to two chloride ions (3.13 and 3.32 Å), while N$_2$ (planar form) has two ordinary bonds N$_2$−H...Cl (3.32), N$_2$−H...O$_4$ (2.93 Å) and one bifurcated bond to O$_2$ and O$_4$ (3.04 and 2.98 Å).

There are also some short C−−−O and C−−−Cl distances, which evidently relate to weak hydrogen bonds: C−H...O, C−H...Cl.

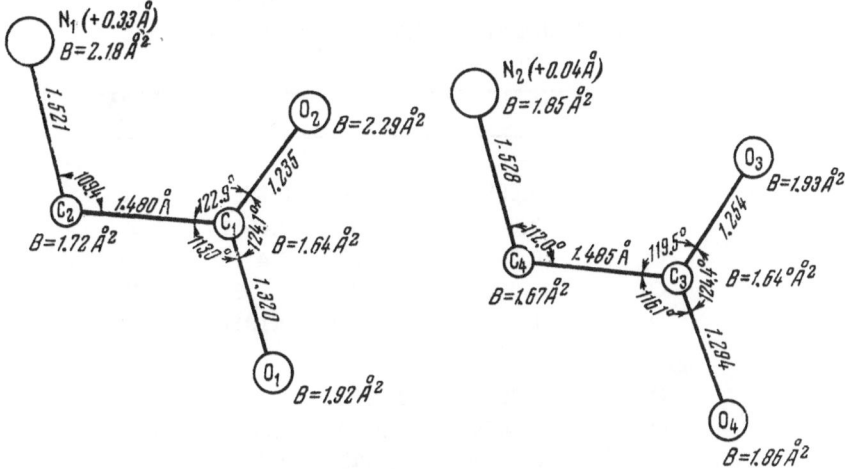

Fig. 20. Molecular form in the structure of diglycine hydrochloride.

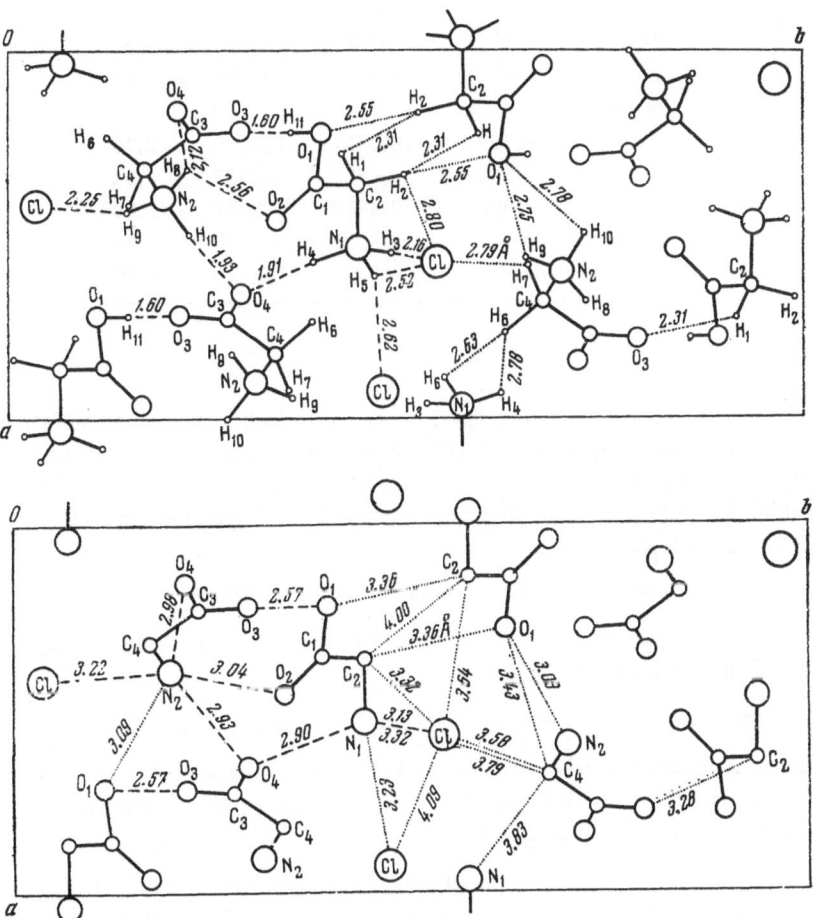

Fig. 21. Schematic representation of the structure of diglycine hydrochloride,
view along c axis.

TABLE 14. Parameters of the Main Atoms in the Structure of Triglycine Sulfate (Temperature Factors of the Form $T_j = \exp - B_j \sin^2 \theta / \lambda^2$)

Atom	x	y	z	B_j, $\mathring{A}^2$
Sulfate ion				
S	0.9995	0.2500	0.2250	0.33
O_1	0.8583	0.2447	0.0051	1.56
O_2	0.9669	0.2437	0.4572	1.83
O_3	1.0920	0.1565	0.2234	1.48
O_4	1.0769	0.3469	0.1941	2.27
gly I				
O	0.6064	0.2393	1.0746	2.66
O'	0.4935	0.2718	0.6668	3.39
C	0.4905	0.2472	0.8727	2.75
C^N	0.3348	0.2361	0.9049	2.78
N	0.3595	0.2110	1.1639	3.86
gly II				
O	0.2218	0.4975	0.7646	1.79
O'	0.4596	0.5397	0.7988	3.23
C	0.3153	0.5331	0.6797	2.82
C^N	0.2675	0.5734	0.4070	3.29
N	0.0939	0.5800	0.3063	2.39
gly III				
O	0.7824	0.4931	0.2229	2.76
O'	0.5454	0.4825	0.2317	2.57
C	0.6937	0.4749	0.3281	1.32
C^N	0.7440	0.4320	0.5906	1.27
N	0.9068	0.4331	0.7059	1.06

Marked ferroelectric behavior occurs in some glycine salts, such as triglycine sulfate $3gly \cdot H_2SO_4$ [45, 46] and the isomorphous compounds $3gly \cdot H_2SeO_4$ [47, 48] and $3gly \cdot H_2BeF_4$ [46], as well as in $gly \cdot MnCl_2 \cdot 2H_2O$ [49] and $gly \cdot AgNO_3$ [50]. A complete structure study has been performed only for the first [46], whose ferroelectric phase is stable up to 47°C and which has space group $P2_1$, cell parameters $a = 9.417$ Å; $b = 12.643$ Å; $c = 5.735$ Å; $\beta = 110°23'$; $Z = 2[3 gly \cdot H_2SO_4]$.

The isomorphism with the selenate was used in deducing the structure.

The structure was refined by three-dimensional least squares, the final $R(hkl)$ being 16.7%. Table 14 gives the parameters of the main atoms.

Figure 22 gives the stereochemical characteristics of the sulfate ions and glycine molecules.

The sulfate ion is a distorted tetrahedron with S—O distances ranging from 1.467 to 1.481 Å (Fig. 22a). The mean S—O distance agrees well with the values found for other organic compounds and is somewhat less than the analogous distance (1.51 Å) in inorganic sulfates.

The glycine occurs as planar positive ions $^+NH_3CH_2COOH$ (gly I and gly II in Fig. 22b) and as zwitterions $^+NH_3CH_2COO^-$ (gly III in Fig. 22b), in which the nitrogen atom lies 0.26 Å from the plane of the rest of the molecule. The two types of ion thus reverse the configurations found for diglycine hydrobromide and hydrochloride.

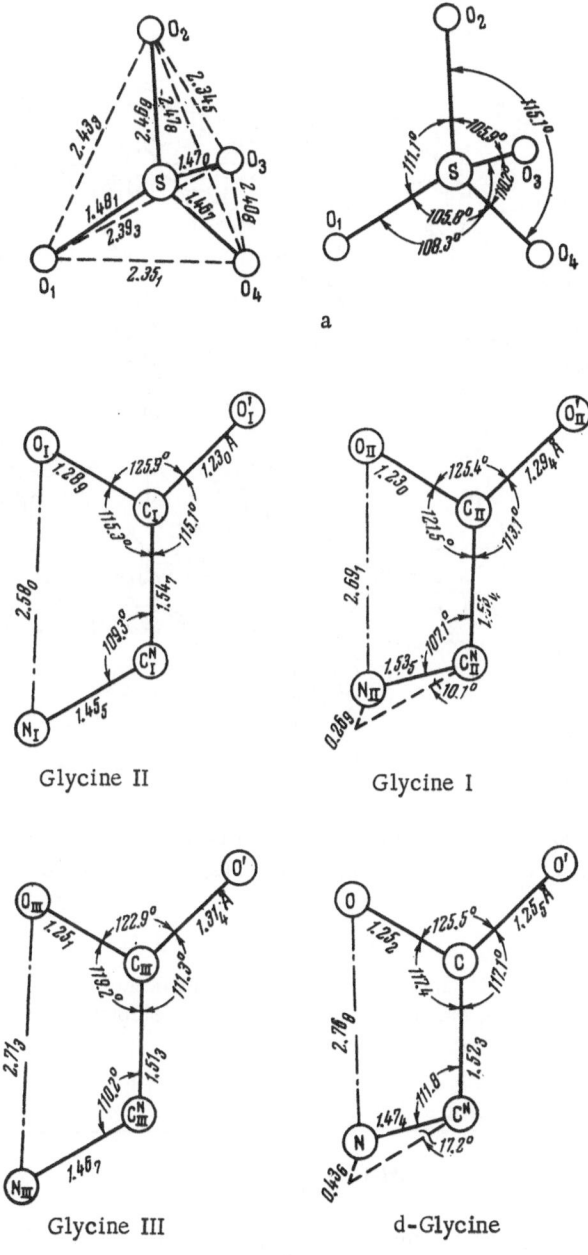

Fig. 22. Structure of (a) sulfate ion, (b) glycine molecules in triglycine sulfate.

The distances and angles in the zwitterion agree well with the standard values, except for $C_{II}-N_{II}$ which takes a value close to that for diglycine hydrochloride [44].

The distances and angles in the positive ion are close to those for α-glycine, except that the C—O bonds are unequal.

The hydrogen-bond system is fairly complicated (Fig. 23). Each positive ion I(A) lies near the y=1/4 plane; it has an O—H...O bond between the COOH group and atom O(A) in the sulfate as well as three N—H...O bonds from N(A) to atoms $O_3(A^*)$, $O_I{}'(A^*)$, $O_{II}{}'(B^*)$. Ions of types II and III lie near the plane y=1/2 and are linked by strong (2.438 Å) hydrogen bonds ($N_{II}H_3{}^+$ and $N_{III}H_3{}^+$) each form three hydrogen bonds to the oxygen

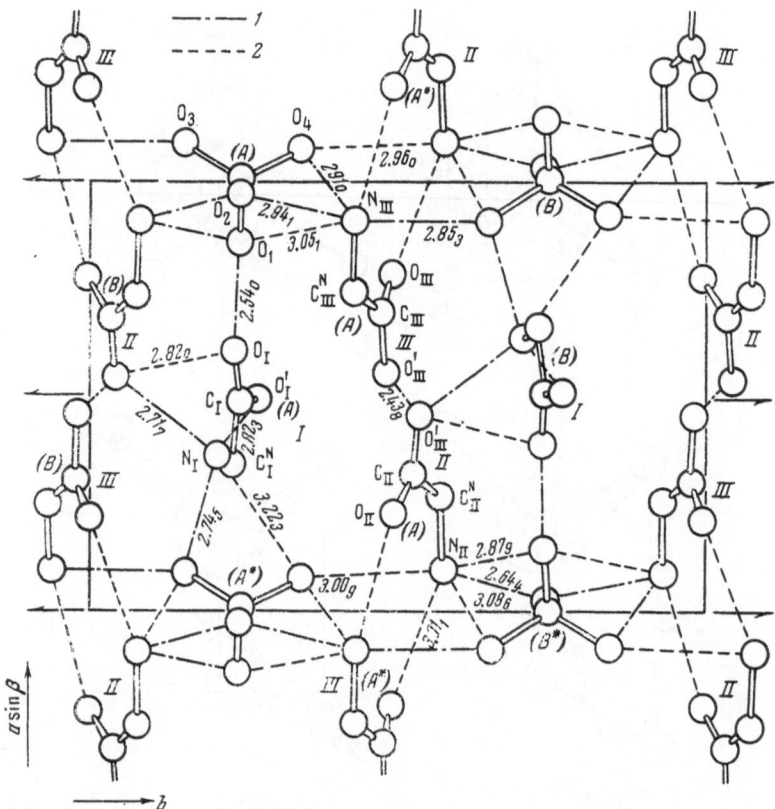

Fig. 23. Schematic representation of the structure of triglycine sulfate,
view along c axis: (1) hydrogen bonds, (2) other close contacts.

atoms of type I ions and sulfate groups. It is supposed that the ferroelectric behavior is due to the very short
O_{III}' (A)−H...O_{II} (A) hydrogen bonds, but further study of the ferroelectric and paraelectric phases is necessary
to elucidate this.

Alanine (α-Aminopropionic Acid)

$CH_3CH(NH_2)COOH$

This was the first amino acid to be synthesized (1850); only later was it found to occur naturally. It
occurs widely, in particular in nearly all proteins, and it is present in especially large amounts in silk fibroin
[18-20].

A complete x-ray study was performed in 1941 [51]; the crystals were grown from an aqueous solution of
the racemate by slow evaporation of the solvent. Space group Pna, cell parameters a = 12.06 Å; b = 6.05 Å;
c = 5.82 Å; ρ_{meas} = 1.40 g/cm^3; Z = 4.

The trial model was derived from the three-dimensional Patterson function with extensive use of Harker
sections, the coordinates of the hydrogen atoms being deduced from crystallographic considerations. The
structure was refined by trial and error. The distances and angles mostly agree with those for α-glycine [24],
but the C−N bond (1.43 Å) is very much shortened.

A subsequent more complete refinement [52] was based on the three-dimensional electron-density distribution: this second refinement reduced the R(hkl) from the complete set of intensities from 15.7 to 14.6%, while

TABLE 15. Coordinates of the Main Atoms
in the Structure of DL-Alanine

Atom	x	y	z
C_1	0.1428	0.3154	0.1622
C_2	0.1632	0.2192	0.4027
C_3	0.0908	0.0192	0.4457
N	0.1397	0.3955	0.5762
O_1	0.0897	0.4832	0.1355
O_2	0.1841	0.1985	0.0019

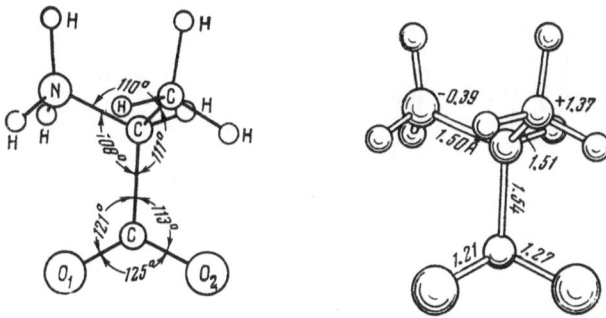

Fig. 24. Molecular structure of DL-alanine.

it gave the distances to ~0.01 Å. Table 15 gives the coordinates of the main atoms. All parameters (Fig. 24) now agreed well with those found for other amino acids. The α-carbon lies in the plane of the COOH group, the nitrogen atom projecting by 0.39 Å and the methyl carbon by 1.37 Å. The COOH group is not symmetrical about the C_1–C_2 bond, because the two oxygen atoms participate differently in the hydrogen-bond system.

There is a network of strong N–H...O hydrogen bonds (Fig. 25). Each nitrogen atom forms three hydrogen bonds to oxygen atoms in the carboxyl groups of adjacent molecules, these bonds having a tetrahedral disposition; this and the absence of O–H...O bonds would indicate that the molecule is present as a zwitterion. The molecular chains parallel to the c axis stand out from the general framework; the alanine molecules are linked sequentially by N–H...O_2 bonds (2.80 Å), while N–H...O_2 (2.84 Å) and N–H...O_1 (2.88 Å) bonds join the chains into tubes, each tube consisting of six chains and each chain being shared by three tubes. The CH_3 groups of diametrically opposite chains are directed within the tube in opposite senses; they cannot rotate, because they are involved in numerous contacts.

The structure of L-alanine has also been determined [53]. The crystals were grown from aqueous solution by slow evaporation; space group $P2_12_12_1$, cell parameters very close to those of the racemic form: a = 6.032 Å; b = 12.343 Å; c = 5.784 Å; Z = 4.

The intensity distribution in the hk0 reflections was also the same as that for the racemic form, so the xy projections must be similar. The initial x and y coordinates were taken as those found [51] for DL-alanine, while the z coordinates were deduced on the basis of the possible packing and the system of hydrogen bonds.

The preliminary refinement was done by least squares neglecting the hydrogen atoms; in this way R(hkl) was reduced to 13%, at which point the hydrogen atoms were included, the positions of these being deduced from a differential Fourier synthesis together with crystallochemical considerations. The final parameters are given in Table 16 and correspond to a R(hkl) of 4.9%; the limits of error for the C–C, C–O, and C–N distances are ~0.004 Å, while those for C–H and N–H are ~0.04 Å.

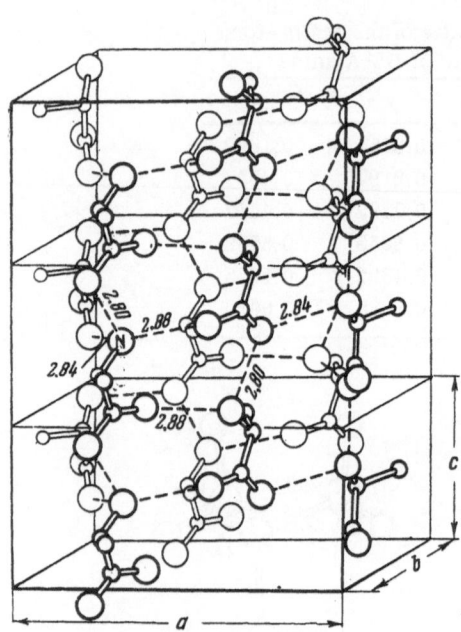

Fig. 25. General form of the structure
of DL-alanine.

TABLE 16. Parameters of the Atoms in the Structure of L-Alanine [53] [Temperature Factors of the Form
$T_j = \exp-(B_{11}h^2 + B_{22}k^2 + B_{33}l^2 + B_{12}hk + B_{13}hl + B_{23}kl)]$

Atom	x	y	z	$B_{11} \cdot 10^4$	$B_{22} \cdot 10^4$	$B_{33} \cdot 10^4$	$B_{12} \cdot 10^4$	$B_{13} \cdot 10^4$	$B_{23} \cdot 10^4$
C_1	0.5606	0.1418	0.6016	149	25	119	-6	-18	-8
C_2	0.4769	0.1612	0.3563	138	28	98	15	5	-4
C_3	0.2746	0.0915	0.3025	177	58	153	-38	-61	-15
N	0.6560	0.1382	0.1856	150	36	88	5	12	2
O_1	0.7287	0.0843	0.6283	206	44	151	59	-47	16
O_2	0.4501	0.1850	0.7609	227	53	91	46	18	-10

(Temperature Factors of the Form $T_j = \exp-B_j \sin^2 \theta/\lambda^2$)

Atom	x	y	z	B_j	Atom	x	y	z	B_j
H_1	0.699	0.064	0.194	1	H_5	0.222	0.108	0.144	3
H_2	0.769	0.179	0.223	1	H_6	0.153	0.103	0.410	3
H_3	0.599	0.150	0.043	1	H_7	0.315	0.014	0.306	3
H_4	0.442	0.239	0.346	1					

The structure is very similar to that of DL-alanine (Figs. 25 and 26). All the protons are involved in
N−H...O bonds, of which the N−H...O_2 ones (length 2.813 Å) join the molecules into infinite chains parallel
to the c axis. In addition, N−H...O_2 (2.829 Å) and N−H...O_1 (2.853 Å) bonds join the chains into a three-
dimensional framework.

The structure of the racemate may be transformed to that of the L-form by converting the chains of D
molecules into L ones by reflection in (001) and displacement along the c axis by the amount needed to produce the N−H...O_2 (2.853 Å) bonds. Both structures have the same projection along the c axis, which leads to
identical distributions for the hk0 intensities.

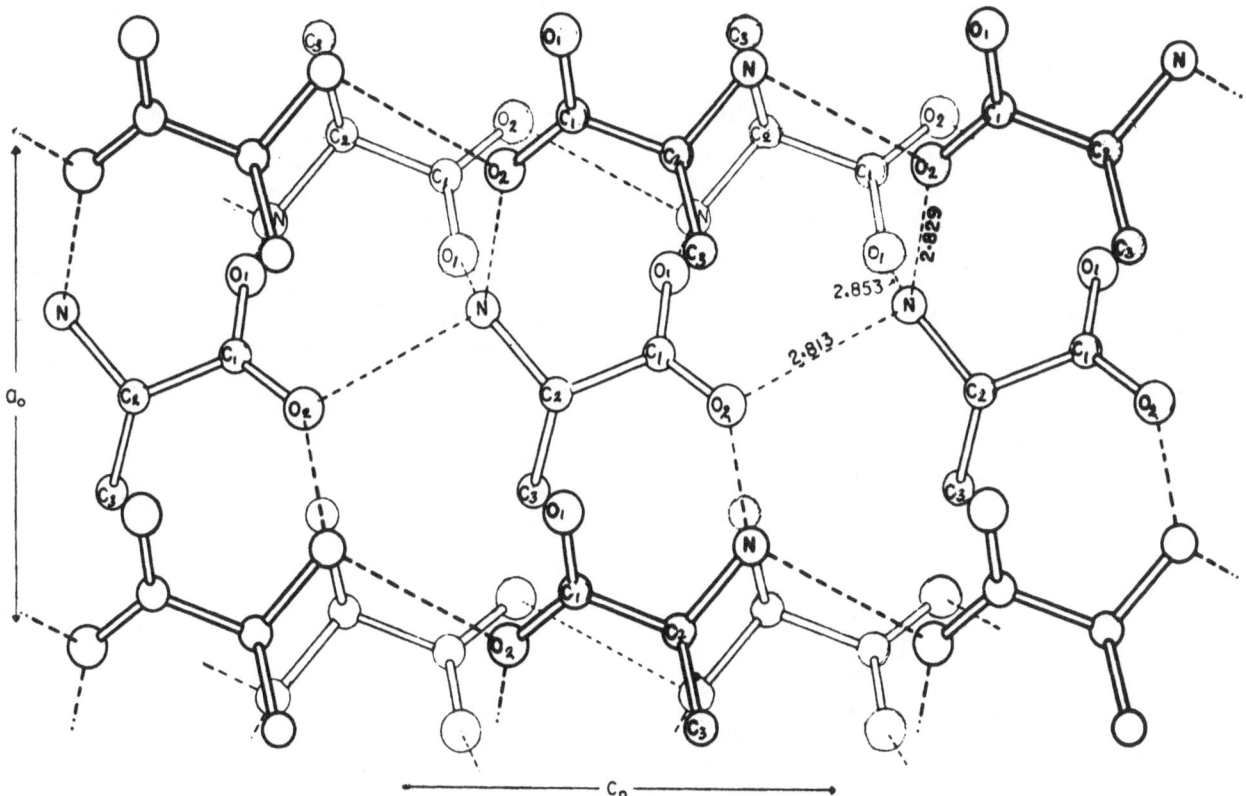

Fig. 26. Schematic representation of the structure of L-alanine as seen along the b axis.

TABLE 17. Bond Lengths Involving Hydrogen
Atoms in the Structure of L-Alanine

$N-H_1$	0.95 Å	C_3-H_5	0.99 Å
$N-H_2$	0.88	C_3-H_6	0.98
$N-H_3$	0.90	C_3-H_7	0.99
C_2-H_4	0.98		

Figure 27a and Table 17 give the bond lengths and valence angles in the molecule of L-alanine. All of these, except the C_1-O_1 bond, agree within the error of measurement with the results [52] for the racemic form. The C_1-O_1 bond is elongated by 0.04 Å, which means that the asymmetry of the carboxyl group is here less pronounced.

Figure 27b shows the configuration. The atoms of the carboxyl group lie in a plane, the nitrogen atom deviating from this plane by 0.450 Å and the carbon atom of the methyl group by 1.384 Å.

A refinement by reference to diffractometer data [54] gave nothing essentially new (Table 18); the slight differences from the results of [53] are ascribed to systematic errors.

Some biological compounds (e.g., pantothenic acid and carnosine) contain β-alanine as well as α-alanine. The structure of β-alanine has been examined on the amino acid itself [57], copper alaninate hexahydrate [55], and nickel alaninate dihydrate [56]. The second of these has space group $P2_1/c$ and cell parameters a = 5.46 Å; b = 7.71 Å; c = 18.11 Å; β = 92°; Z = 2 [$Cu(\beta\text{-ala})_2 \cdot 6H_2O$].

TABLE 18. Characteristics of the Structure of L-Alanine [54] Parameters of the Principal Atoms [Temperature Factors of the Form $T_j = \exp-(B_{11}h^2 + B_{22}k^2 + B_{33}l^2 + B_{12}hk + B_{13}hl + B_{23}kl)$]

Atom	x	y	z	$B_{11} \cdot 10^4$	$B_{22} \cdot 10^4$	$B_{33} \cdot 10^4$	$B_{12} \cdot 10^4$	$B_{13} \cdot 10^4$	$B_{23} \cdot 10^4$
C_1	0.5606	0.1413	0.6023	169	27	142	6	-59	12
C_2	0.4764	0.1611	0.3559	163	36	121	42	-8	-1
C_3	0.2744	0.0919	0.3021	195	62	163	-26	-35	-2
N	0.6565	0.1375	0.1853	176	41	99	14	-13	-3
O_1	0.7278	0.0843	0.6280	217	46	177	62	-67	21
O_2	0.4499	0.1856	0.7604	233	52	118	44	25	-10

Lengths of Intramolecular Bonds

C_1-O_1	1.239 Å	C_2-C_3	1.523 Å
C_1-O_2	1.257	C_2-N	1.496
C_1-C_2	1.533		

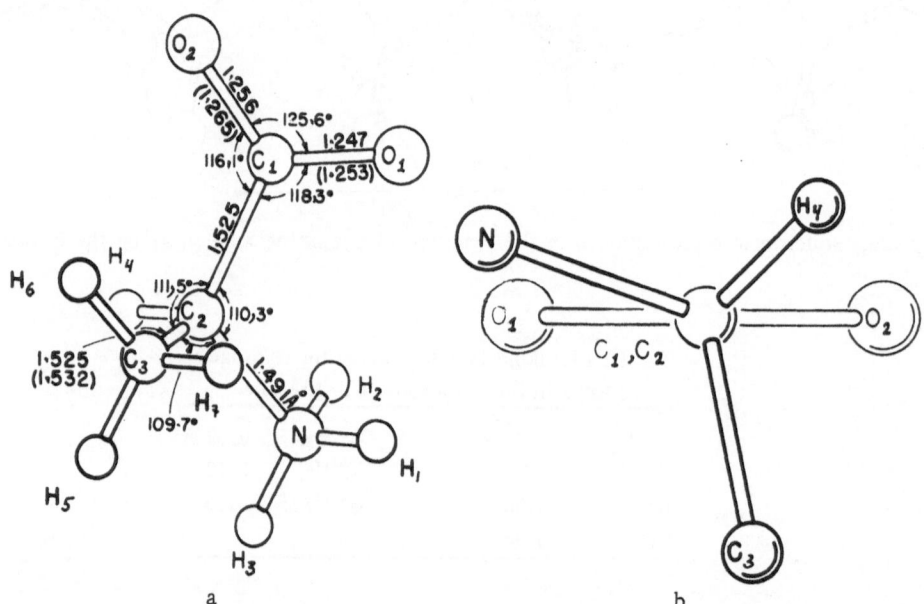

Fig. 27. Molecular form in the structure of L-alanine: (a) bond lengths and valence angles, with the values corrected for thermal motion of the atoms in parentheses; (b) configuration of molecule around C_1-C_2 bond.

The heavy-atom method gave a model of the structure, the coordinates (Table 19) being refined via electron-density projections. The final results gave $R(0kl) = 19\%$, $R(h0l) = 22\%$.

Figure 28 gives a general view of the structure, which consists of distorted octahedra around the copper atoms, each of these containing two alanine molecules and two water molecules. The covalent Cu−O and Cu−N bonds have lengths close to those found [120] for copper prolinate dihydrate.

The distances and angles in β-alanine are close to those found for other amino acids. The NH_2 and COOH groups are in the gauche configuration relative to the C_2-C_3 bond (there is an angle of 70° between the planes formed by the triplets C_1, C_2, C_3 and C_2, C_3, N).

TABLE 19. Coordinates of the Main Atoms in the
Structure of Copper β-Alaninate Hexahydrate

Atom	x	y	z
Cu	0.000	0.000	0.000
C_1	-0.185	-0.025	0.158
C_2	0.065	-0.118	0.185
C_3	0.187	0.228	0.130
O_1	-0.205	0.005	0.095
O_2	-0.325	-0.005	0.206
O_3	0.187	0.280	0.040
O_4	0.450	0.322	0.172
O_5	-0.258	0.445	0.060
N	0.278	-0.135	0.058

TABLE 20. Parameters of the Main Atoms in the Structure of Nickel β-Alaninate Dihydrate
(Temperature Factors of the Form $T_j = \exp{-B_j \sin^2 \theta / \lambda^2}$)

Atom	x	y	z	$B_j(hk0)$, Å^2	$B_j(0kl)$, Å^2	$B_j(h0l)$, Å^2
Ni	0	0	0	1.8	0.5	1.5
N	0.2219	0.0206	0.2428	1.7	1.4	0.9
C_1	0.3675	0.1638	0.1813	2.2	1.2	2.7
C_2	0.3373	0.3878	0.2173	2.9	0.8	2.6
C_3	0.2155	0.4038	-0.0247	0.9	0.8	1.0
O_1	0.1095	0.2555	-0.1631	2.1	1.2	2.1
O_2	0.2456	0.5811	-0.0847	2.3	1.3	3.2
O_3	-0.0690	0.2176	0.3355	2.7	1.0	3.3

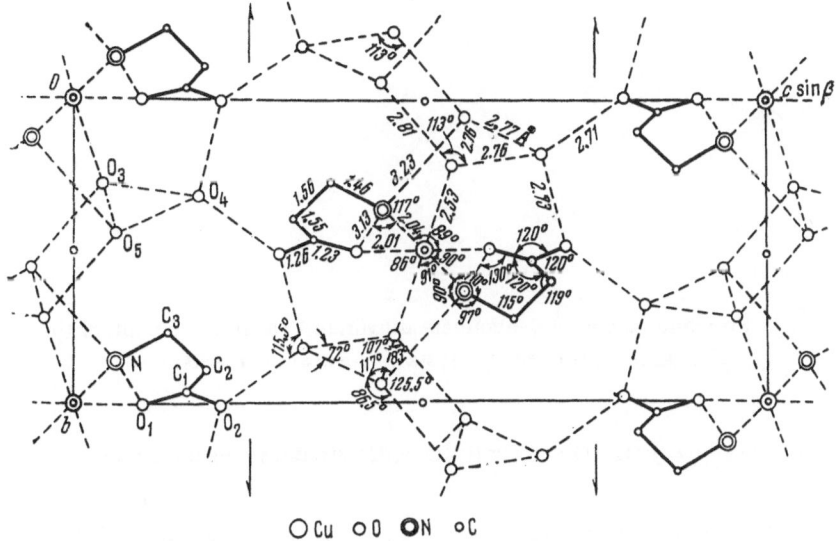

○ Cu   ○ O   ● N   ○ C

Fig. 28. Structure of copper β-alaninate hexahydrate, view along a axis.

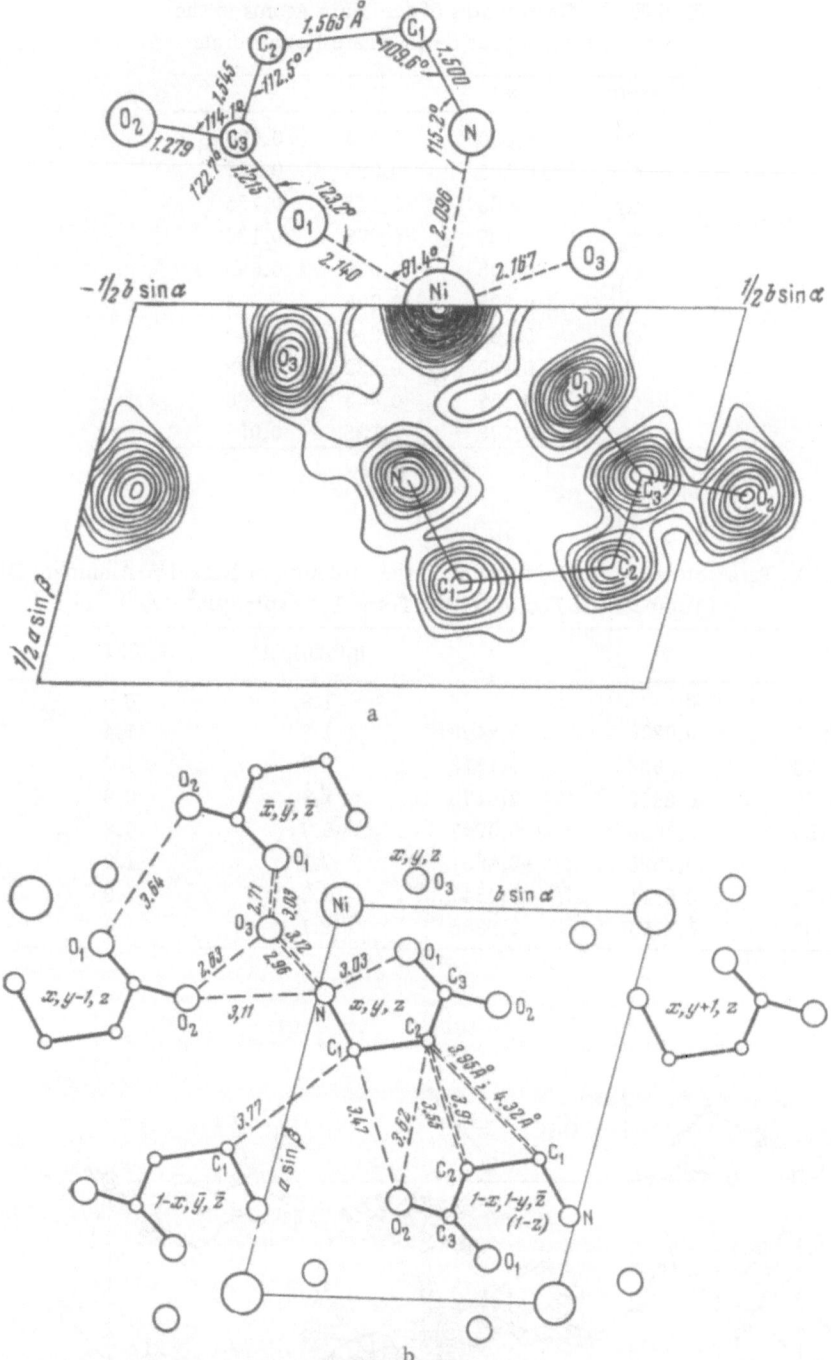

Fig. 29. Structure of nickel β-alaninate dihydrate: (a) single complex, (b)
linking of complexes in crystal. View along c axis.

The octahedra are linked by N−H...O and O−H...O bonds via the water molecules not included in the octahedra.

The above nickel salt of β-alanine has space group $P\bar{1}$ and cell parameters $a = 8.48$ Å; $b = 6.77$ Å; $c = 4.93$ Å; $\alpha = 103.0°$; $\beta = 95.2°$; $\gamma = 102.3°$; $\rho_{meas} = 1.720$ g/cm³; $Z = 1$ [Ni(β-ala)$_2$ · 2H$_2$O]; $\rho_X = 1.689$ g/cm³.

The model was derived from Patterson projections, while the structure was refined via differential electron-density projections and two-dimensional least-squares treatments. The final R were 15% for hk0, 12.4% for 0kl, and 15.3% for h0l. The accuracies are: C$-$X (C, N, O) $\sim$0.04 Å, Ni$-$X(O, N) $\sim$0.02 Å, valence angles $\sim$2°. Table 20 gives the parameters of the main atoms.

The bond lengths and angles (Fig. 29) are close to the standard values.

The nickel salt resembles the copper one in consisting of octahedral complexes; the vertices of a complex are the oxygen and nitrogen atoms of the two β-alanine residues and the two water molecules. The line of the $O_3(H_2O)-Ni-O_3(H_2O)$ bonds lies at 88.2° to the plane of the atoms in the amino acid residues, which themselves are not planar. There is an angle of 30.3° between the planes passing through O_1, O_2, C_2, C_3 and C_1, C_2, C_3, and also one of 73.7° between the planes through C_1, C_2, C_3 and C_1, C_2, N.

The complexes are linked together by N$-$H...O and O$-$H...O bonds; they form layers parallel to the (100) planes (Fig. 29b); whose disposition perpendicular to the (100) planes is controlled by intermolecular forces.

Crystals of β-alanine itself were grown from aqueous solution: a = 9.865 Å; b = 13.81 Å; c = 6.07 Å; Z = 8.

The model was derived from three Harker lines: $P(x, \frac{1}{2}, 0)$, $P(0, y, \frac{1}{2})$, and $P(\frac{1}{2}, 0, z)$.

The structure was refined via electron-density projections and by three-dimensional least squares; R(hk$l$) was 15.7%. The distances were determined to $\sim$0.010 Å and the valence angles to $\sim$0.6°. Table 21 gives the coordinates of the main atoms.

Figure 30 shows the bond lengths and valence angles of β-alanine.

It is of interest to compare the molecular configurations in β-alanine and in the above salts.

In every case atoms C_2, C_3, O_1, and O_2 are coplanar, this plane lying at 9.3° to the plane of C_1, C_2, C_3 in the case of β-alanine and at 30.3° in the case of the nickel salt. There is less difference in the angles between the plane of C_1, C_2, C_3 and that of C_1, C_2, N, namely 83.8° in β-alanine, 73.7° in the nickel salt, and 70° in the copper salt.

Figure 30 shows the packing of the molecules in the crystal. The molecules form layers parallel to the (010) planes and are linked within the layers via N$-$H...O_1 ($\frac{1}{2}+x$, y, $-\frac{1}{2}-z$) and N$-$H...O_2 ($\frac{1}{2}+x$, y, $\frac{1}{2}-z$) hydrogen bonds, while adjacent layers are held together by N$-$H...O_1($\bar{x}$, $\bar{y}$, $\bar{z}$) hydrogen bonds and van der Waals forces.

Valine (α-Aminoisovalerianic Acid)

$(CH_3)_2CHCH(NH_2)COOH$

Valine occurs in many proteins, but usually in small amounts; structures have been reported for L-valine· HBr [58], L-valine HCl [59], and L-valine HCl monohydrate [60]. The space group and cell parameters are known also for the L and DL forms of the amino acid [61, 62].

TABLE 21. Coordinates of the Main Atoms
in the Structure of β-Alanine

Atom	x	y	z
N	0.2955	0.0663	-0.0450
C_1	0.2318	0.1599	0.0120
C_2	0.0850	0.1601	-0.0796
C_3	-0.0214	0.1107	0.0717
O_1	-0.1393	0.0969	-0.0156
O_2	0.0174	0.0851	0.2655

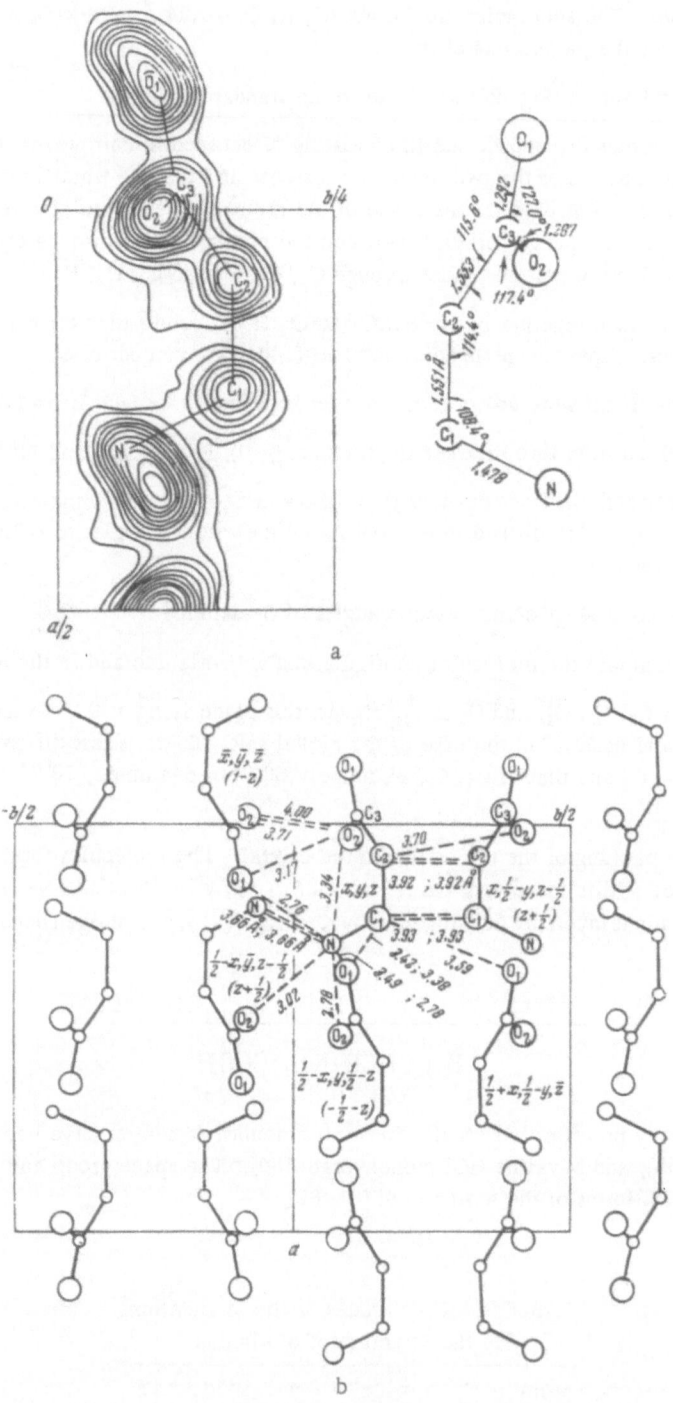

Fig. 30. Structure of β-alanine: (a) molecule, (b) packing of molecules in crystal, seen along c axis; — — —represent the principal intermolecular distances.

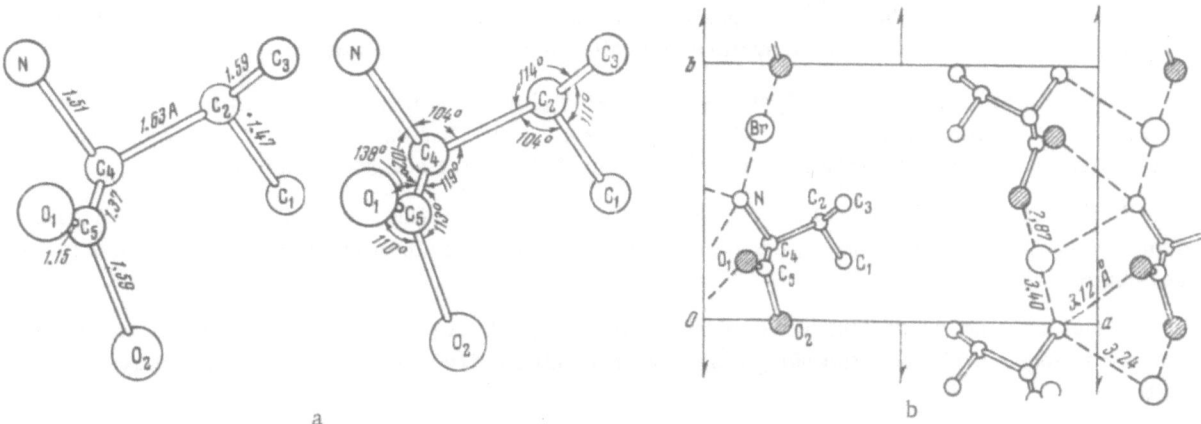

Fig. 31. Structure of L-valine HBr: (a) L-valine molecule, (b) packing of molecules in crystal; view along c axis.

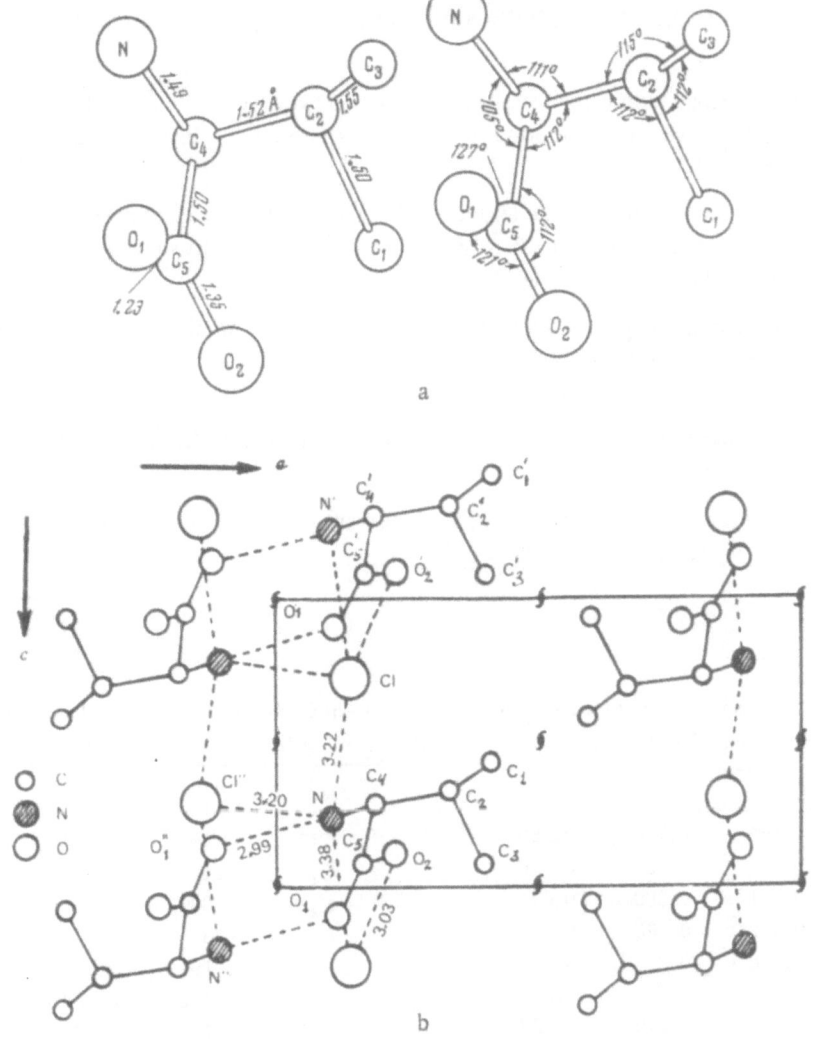

Fig. 32. Structure of L-valine HCl: (a) molecular structure, (b) packing
in crystal; view along c axis.

The data on L-valine have not been published, nor have those given below for L-leucine HBr and ·L-arginine HCl; they have kindly been made available by colleagues at Madras University.

L-valine H Br has space group $P2_1$ and cell parameters [58] of a = 10.18 Å; b = 7.34 Å; c = 5.55 Å; β = 90°15'; Z = 2. The structure has been established by Partasarati and Chandrasekaren. The refinement was by least squares with individual isotropic temperature factors (Table 22); R(h0l) and R(hk0) were 12%.

Figure 31a shows the distances and angles in the molecule of L-valine, of which C_2-C_4 and C_5-O_2 are lengthened and C_4-C_5 is shortened. These deviations appear to arise from failure to refine with respect to the three-dimensional set of intensities.

Figure 31b shows the packing of the molecules in the crystal. A valine molecule is linked to others and to Br^- by hydrogen bonds, the result being double layers parallel to (100), which are held together by van der Waals forces.

Crystals of L-valine HCl [59] are isomorphous with those of the hydrobromide: space group $P2_1$, cell parameters a = 10.38 Å; b = 7.08 Å; c = 5.46 Å; β = 91°30'; Z = 2. This structure was determined by Partasarati and Ramachandran with fairly high accuracy; R(hk$l$) was 13% with the use of anisotropic temperature factors. Table 23 gives the coordinates, while Fig. 32a shows the bond lengths and valence angles, which agree within the limits of determination (~ 0.02 Å and 1°30') with mean values found for structure of other amino acids.

$$..Cl (3.20 \overset{\bullet}{A})$$

The packing is analogous to that for L-valine HBr, with N−H...Cl^- (3.22 Å; 3.38 Å); N−H...O (2.97 Å); and O_2-H...Cl^- (3.03 Å) hydrogen bonds joining the molecules into double layers, which are held together by van der Waals forces (Fig. 32b).

Thyagaraja Rao [60] has determined the structure of L-valine HCl monohydrate: space group $P2_12_12_1$, cell parameters a = 6.85 Å; b = 21.20 Å; c = 6.17 Å; Z = 4. Refinement by least squares with individual isotropic temperature factors gave R(hk$l$) as 11% (Table 18).

Figure 33a shows the bond lengths and valence angles; Fig. 33b shows the packing. The valine molecules are linked by hydrogen bonds via the Cl^- ions and water molecules to form double layers parallel to (010), which are held together by van der Waals forces.

DL-valine [61] has space group P1 or P$\bar{1}$, with cell parameters of a = 5.25 Å; b = 5.43 Å; c = 11.05 Å; α = 91°; β = 92.4°; γ = 109.4°; Z = 2.

L-valine [62] has space group $P2_1$, with cell parameters of a = 9.71 Å; b = 5.32 Å; c = 12.08 Å; β = 90.8°; ρ_{meas} = 1.230 g/cm³; Z = 4; ρ_{meas} = 1.247 g/cm³.

TABLE 22. Coordinates of the Main Atoms in the Structure of L-Valine HBr			
Atom	x	y	z
Br^-	0.164	0.750	0.289
C_1	0.415	0.238	0.566
C_2	0.341	0.389	0.674
C_3	0.412	0.463	0.910
C_4	0.195	0.304	0.724
C_5	0.174	0.204	0.929
N	0.110	0.470	0.767
O_1	0.126	0.227	0.116
O_2	0.223	0.999	0.906

TABLE 23. Coordinates of the Main Atoms in the Structure of L-Valine HCl			
Atom	x	y	z
Cl^-	0.149	0.750	0.283
C_1	0.409	0.228	0.572
C_2	0.331	0.384	0.683
C_3	0.398	0.470	0.914
C_4	0.191	0.327	0.715
C_5	0.175	0.199	0.930
N	0.109	0.495	0.764
O_1	0.118	0.234	0.120
O_2	0.229	0.028	0.895

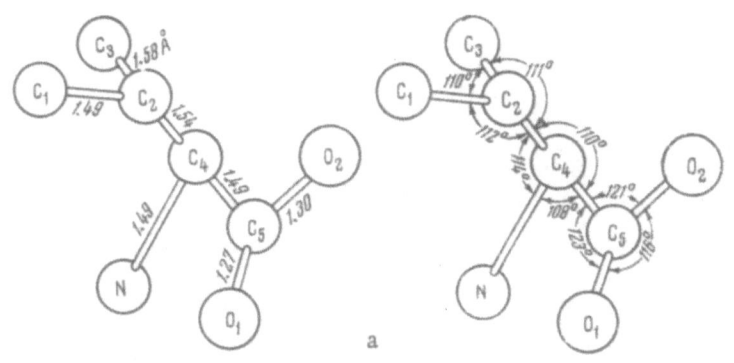

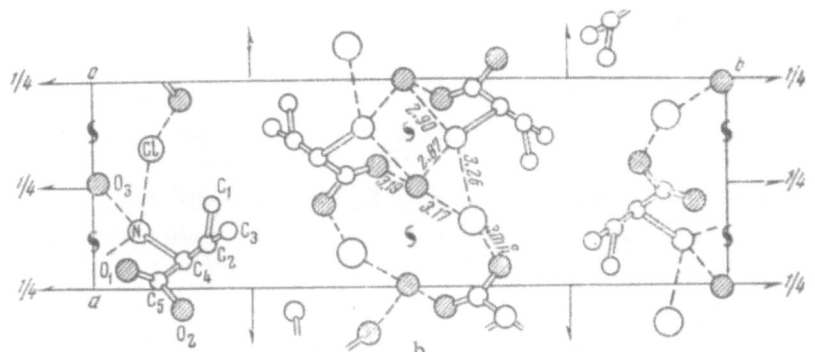

Fig. 33. Structure of L-valine HCl monohydrate: (a) bond lengths and valence angles in molecule of L-valine; (b) packing of molecules in crystal; view along c axis.

Norvaline $CH_3CH_2CH_2CH(NH_2)COOH$ has an unbranched hydrocarbon chain; it is not very common. It seems that it resembles norleucine in occurring in several modifications. The space group and cell parameters have been determined [66] only for the β form of DL-norvaline: I2/a, cell parameters a = 9.93 Å; b = 4.78 Å; c = 30.04 Å; β = 100°; ρ_{meas} = 1.11 g/cm³; Z = 8.

Leucine (α-Aminoisocaproic Acid)
$(CH_3)_2CHCH_2CH(NH_2)COOH$

This is one of the most common amino acids and occurs in many proteins in large amounts [18-20].

The cell parameters and space group have been determined for the racemic mixture [61] and for D-leucine [63], while the structure of L-leucine HBr has been established [64].

DL-leucine [61] has space group P1, with cell parameters a = 5.39 Å; b = 14.6 Å; c = 5.18 Å; α = 103°; β = 111.5°; γ = 96°; ρ_x = 1.18 g/cm³; Z = 2.

D-leucine [63] has space group $P2_122_1$, with cell parameters a = 5.36 Å; b = 14.7 Å; c = 9.65 Å; Z = 4. Subramanian's model was refined by least squares with individual isotropic temperature factors, the final R(hkl) being 16%. Table 19 gives the atomic parameters.

Figure 34 shows the distances and angles implied by these coordinates, which are correct to ~ 0.06 Å and 5°; within these limits they are close to those characteristic of other amino acids.

Figure 34b shows the packing of the molecules in the crystal. Hydrogen bonds N−H...Br⁻ (3.26 Å; 3.41 Å), N−H...O$_1$ (2.82 Å), and O$_2$−H...Br⁻ (3.17 Å) join the molecules into double layers parallel to (010), which themselves are held together by van der Waals forces.

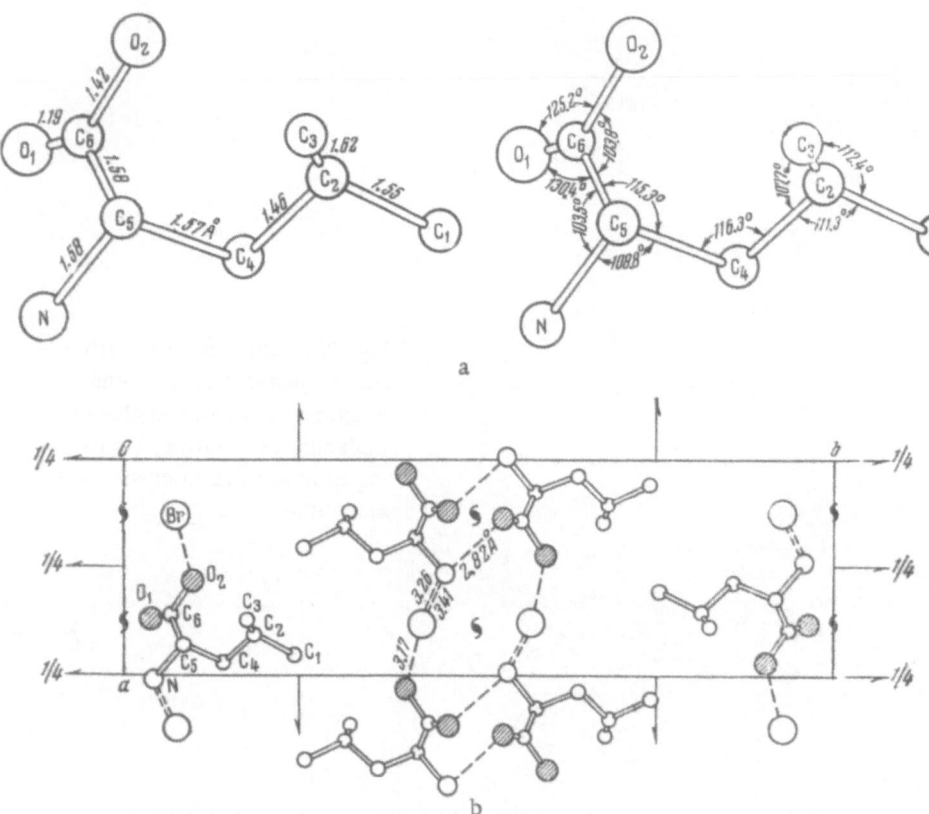

Fig. 34. Structure of L-leucine HBr: (a) molecule of L-leucine, (b) packing of molecules in the crystal; view along c axis.

TABLE 24. Atomic Coordinates in the Structure of L-Valine HCl

Atom	x	y	z
Cl⁻	0.3220	0.0953	0.1281
C_1	0.6016	0.1888	0.6278
C_2	0.7223	0.2121	0.0080
C_3	0.7756	0.1873	0.7734
C_4	0.8682	0.1208	0.7889
N	0.7177	0.0710	0.8464
C_5	0.9556	0.1035	0.5791
O_1	0.9048	0.0556	0.4683
O_2	0.0983	0.1402	0.5194
$O_3(H_2O)$	0.4841	0.0129	0.5169

TABLE 25. Parameters of the Main Atoms in the Structure of L-Leucine HBr (Temperature Factors of the Form $T_j = \exp - B_j \sin^2\theta/\lambda^2$)

Atom	x	y	z	$B_j, \text{Å}^2$
Br⁻	0.2480	0.0741	0.0799	4.18
O_1	0.7339	0.0386	0.2503	3.58
O_2	0.5529	0.0967	0.4766	4.28
N	0.0167	0.0421	0.5961	2.97
C_1	0.8914	0.2436	0.6180	4.22
C_2	0.8019	0.1872	0.6567	3.95
C_3	0.7295	0.1791	0.9277	7.39
C_4	0.9298	0.1429	0.6031	2.65
C_5	0.8528	0.0837	0.6262	3.26
C_6	0.7128	0.0658	0.4230	2.87

TABLE 26. Coordinates of the Principal Atoms

	D(-)Isoleucine · HCl · H₂O				D(-)Isoleucine · HBr · H₂O		
Atom	-x	y	z	Atom	-x	y	z
C_1	0.708	0.158	0.692	C_1	0.733	0.156	0.678
C_2	0.487	0.148	0.600	C_2	0.528	0.144	0.575
C_3	0.503	0.090	0.522	C_3	0.517	0.087	0.478
C_4	0.628	0.090	0.328	C_4	0.542	0.039	0.622
C_5	0.694	0.038	0.231	C_5	0.350	0.033	0.761
C_6	0.276	0.065	0.489	C_6	0.692	0.092	0.325
O_1	0.758	0.129	0.839	O_1	0.761	0.122	0.825
O_2	0.799	0.203	0.660	O_2	0.856	0.194	0.650
N	0.417	0.190	0.467	N	0.467	0.192	0.467
Cl^-	0.153	0.168	0.072	Br^-	0.186	0.161	0.072
$O(H_2O)$	0.756	0.243	0.226	$O(H_2O)$	0.794	0.242	0.242

Isoleucine (α-Amino-β-Methylvalerianic Acid)

$$CH_3CH_2CHCH(NH_2)COOH$$
$$|$$
$$CH_3$$

This occurs, usually with leucine, in most proteins [18-20]; it was isolated from a protein hydrolyzate by Erlich in 1904. The molecule contains two asymmetric carbon atoms, so there are four isomers. The spatial configuration of D(-)isoleucine has been established by x-ray methods [65].

Crystals suitable for x-ray use were produced by adding concentrated HCl and HBr to a suspension of D(-)isoleucine in acetone, with subsequent slow evaporation of the solvent. These were the monohydrates of isoleucine HCl and HBr, whose characteristics were as follows.

D(-)-isoleucine · HCl · H₂O: Space group $P2_12_12_1$, cell parameters a = 6.13 Å; b = 25.01 Å; c = 6.79 Å; ρ_{meas} = 1.19 g/cm³; Z = 4 (ileu · HCl · H₂O); ρ_{meas} = 1.18 g/cm³.

D(-)-isoleucine · HBr · H₂O: Space group $P2_12_12_1$, cell parameters a = 6.21 Å; b = 24.40 Å; c = 7.00 Å; Z = 4 (ileu · HBr · H₂O).

Although the two have similar cell parameters, they are not isomorphous.

A trial model for the first was deduced from (yz) and (yx) Patterson projections by the method of vector convergence. Refinement was by electron-density projection, differential synthesis, and finally two-dimensional least squares (Table 26). The R were 14.7% for 0kl and 14.8% for hk0.

The hydrobromide structure was deduced by the heavy-atom method and was refined via successive electron-density projections (Table 26); the final results gave R(0kl) = 14.5%, R(hk0) = 14.1%.

Anomalous scattering of U $L\alpha_1$ by the Br atoms was also used to establish the absolute configuration of D(-)-isoleucine.

Table 27 and Fig. 35 give the distances and angles for both compounds.

Assuming that the distances were found to 0.05 Å and the angles to 10°, the results do not deviate from standard values.

TABLE 27. Interatomic Distances

D(-)Isoleucine · HCl · H₂O				D(-)Isoleucine · HBr · H₂O			
C_1-O_1	1.27 Å	C_2-C_3	1.55 Å	C_1-O_1	1.33 Å	C_2-C_3	1.54 Å
C_1-O_2	1.27	C_3-C_4	1.53	C_1-O_2	1.22	C_3-C_4	1.56
C_2-N	1.43	C_4-C_5	1.52	C_2-N	1.45	C_4-C_5	1.54
C_1-C_2	1.51	C_5-C_6	1.54	C_1-C_2	1.51	C_5-C_6	1.53

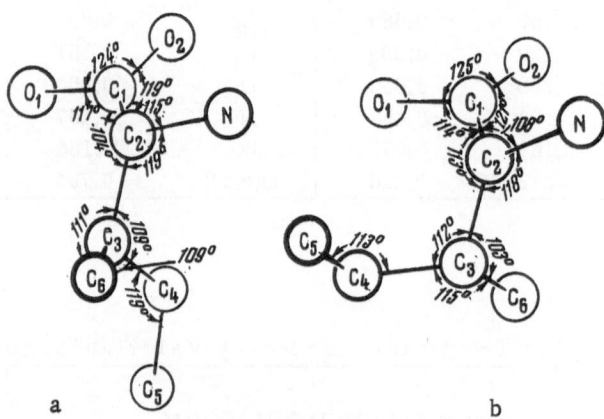

a   b

Fig. 35. Molecular structures in: (a) D(-)-isoleucine ·
HCl · H₂O; (b) D(-)-isoleucine · HBr · H₂O.

The molecular configurations in the two compounds have the following features:

1) The NH_2 and CH_2 groups are relatively cis, which agrees with the chemical evidence;

2) The $NC_2C_1O_1O_2$ groups are planar within the error of the measurements;

3) The $C_1 - - -C_5$ chains have different structures. The hydrochloride has two coplanar groups: C_1, C_2, C_3, C_6 and C_2, C_3, C_4, C_5, while the hydrobromide has only one: C_3, C_4, C_5, C_6. This difference explains why the two are not isomorphous. *

Both compounds contain rigid double layers parallel to (010) (Fig. 36), the molecules within a double layer being linked by $N-H...O$, $N-H...Cl^-(Br^-)$, and $O-H...Cl^-(Br^-)$ hydrogen bonds. There are $N-H...O$ bonds between the amino group and the water molecule, whose lengths are 2.85 and 2.96 Å (2.82 and 2.84 Å).† Tetrahedrally placed with respect to these are $N-H...Cl^-$ (Br^-) [3.18 Å (3.35 Å)]. The $O-H...Cl^-(Br^-)$ bonds are between the carboxyl or water oxygen and the chlorine (bromine). The lengths of the $O(H_2O)-H...Cl^-(Br^-)$ bonds are 3.07 and 3.24 Å (3.30; 3.35 Å), while those of the $O_I-H...Cl^-(Br^-)$ bonds are 3.05 Å (3.29 Å). One double layer is joined to another only by intermolecular forces, which explains the good cleavage on (010).

Norleucine (α-Aminocaproic Acid)

$CH_3CH_2CH_2CH_2CH(NH_2)COOH$

This is a very rare amino acid and occurs mainly in the proteins of nerve tissue [18-20].

---

*Crystals of the hydrochloride isomorphous with those of the hydrobromide were later obtained from a solution in 96% alcohol.

† The quantities in parentheses are the values for D(-)-isoleucine · HBr · H₂O.

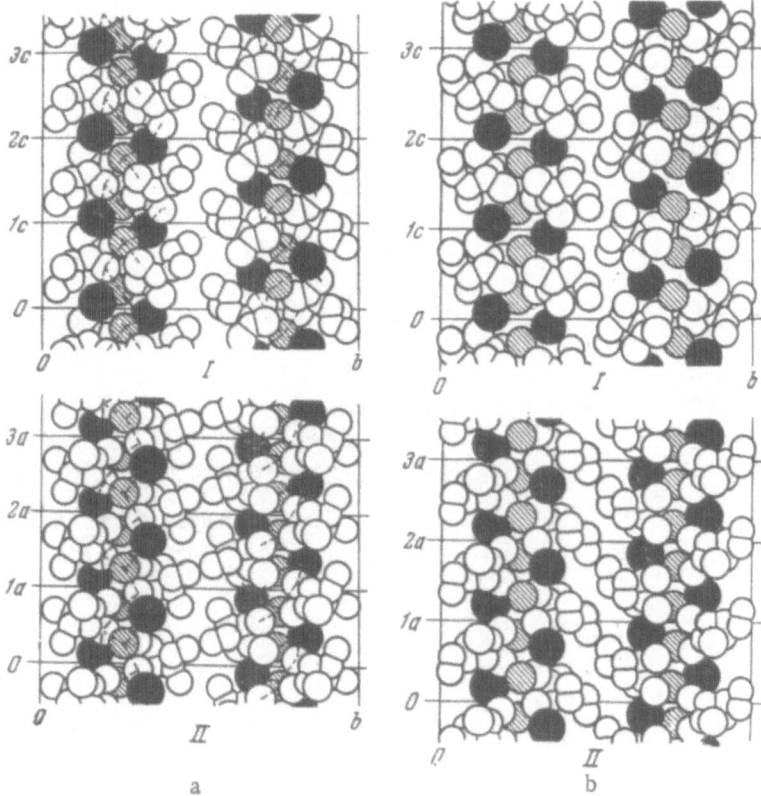

Fig. 36. Structures of (a) D(-)-isoleucine $\cdot$ HCl $\cdot$ H_2O; (b) D(-)-isoleu-
cine $\cdot$ HBr $\cdot$ H_2O; (I) view along a axis; (II) view along c axis.

The crystal structure was examined in 1953 [66] on specimens grown from the racemate in aqueous alcohol by slow evaporation, which included crystals of two species. One of these, the α form, has space group $P2_1/a$ with cell parameters a = 9.84 Å; b = 4.74 Å; c = 16.56 Å; β = 104.5°; ρ_{meas} = 1.16 g/cm^3; Z = 4.

The crystals of the second species were intermediate between this α form and a further β form, which has space group I2/a, with a = 9.84 Å; b = 4.74 Å; c = 33.12 Å; β = 104.5°; ρ_{meas} = 1.16 g/cm^3; Z = 8. Crystals of the pure β form could not be produced.

The model for the structure of the α form was deduced on the basis of the possible resemblance to methionine [103, 104], because the unit cells are similar and the chemical formulas are analogous:

$$CH_3CH_2CH_2CH_2CH(NH_2)COOH \quad \text{(norleucine)},$$

$$CH_3SCH_2CH_2CH(NH_2)COOH \quad \text{(methionine)}.$$

The structure was refined by successive electron-density projections, the final R being 22% for h0l and 24% for 0kl, while the limits were ~0.04 A and ~5°. Table 28 gives the coordinates of the main atoms.

The main interatomic distances and valence angles in α-DL-norleucine are given in Table 29; they are close to those found for methionine. There is some asymmetry in the carboxyl groups and an alternation in the C$-$C bond lengths along the chain.

TABLE 28. Coordinates of the Main Atoms in the Structure of α-DL-norleucine

Atom	x	y	z
O_1	-0.012	-0.067	0.375
O_2	0.169	-0.317	0.432
N	0.356	0.050	0.404
C_1	0.118	-0.114	0.392
C_2	0.209	0.050	0.350
C_3	0.207	-0.069	0.261
C_4	0.301	0.089	0.217
C_5	0.288	-0.036	0.128
C_6	0.394	0.036	0.083

TABLE 29. Interatomic Distances and Valence Angles in the Structure of α-DL-norleucine

C_6-C_5	1.48 Å		$C_6-C_5-C_4$	120°
C_5-C_4	1.57		$C_5-C_4-C_3$	110
C_4-C_3	1.51		$C_4-C_3-C_2$	114
C_3-C_2	1.58		$C_3-C_2-C_1$	112
C_2-C_1	1.49		C_3-C_2-N	110
C_2-N	1.50		C_1-C_2-N	109
C_1-O_1	1.26		$C_2-C_1-O_1$	120
C_1-O_2	1.20		$C_2-C_1-O_2$	118
			$O_1-C_1-O_2$	122
$N(A)...O_1(B)$	2.73 Å		$C_2-N-O_1(B)$	109°
$N(A)...O_1(B')$	2.87		$C_2-N-O_1(B')$	108
$N(A)...O_2(C)$	2.87		$C_2-N-O_2(C)$	105
			$O_1(B)-N-O_1(B')$	116

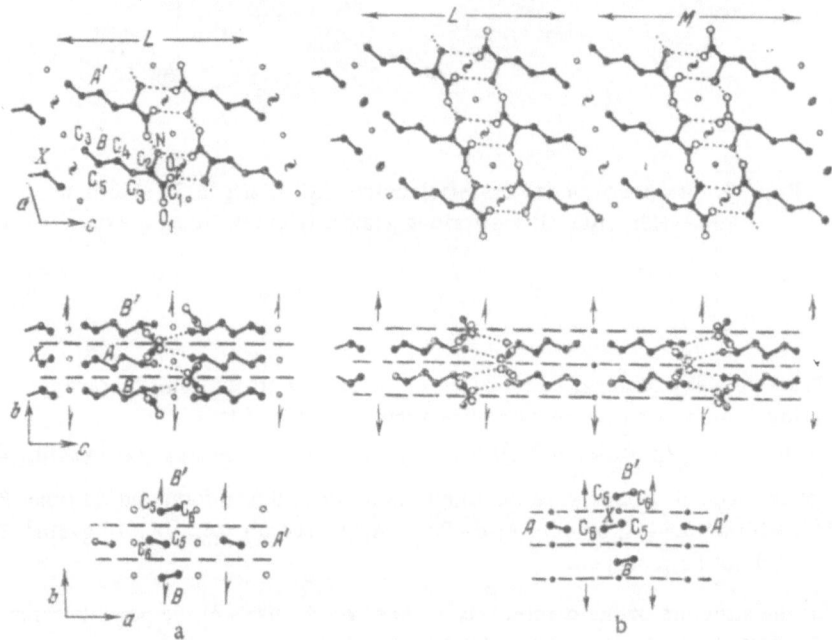

Fig. 37. Schematic representation of the structures of: (a) α-DL-norleucine, (b) β-DL-norleucine.

A complete system of N−H...O hydrogen bonds is formed. The molecules are joined by these bonds into double layers parallel to (001); the double layers are transformed one to another by twofold screw axes and centers of symmetry. There are only intermolecular forces between layers (Fig. 37a).

The β form must consist of the same double layers, but with these related via twofold axes and centers of symmetry, with a different relative packing of the $-CH_2-CH_3$ end groups (Fig. 37b) [66].

2. HYDROXYMONOAMINOMONOCARBOXYLIC ACIDS

Serine (α-Amino-β-Hydroxypropionic Acid)

$HOCH_2CH(NH_2)COOH$

Serine occurs in most proteins; it was first isolated from silk in 1865, and its chemical structure was established in 1902 [18-20].

The crystal structure was first studied in 1942 [67], but only the space group and cell parameters were determined; the complete structure was determined in 1953 [68].

The crystals were grown from an aqueous solution of the racemate by slow evaporation. The space group is $P2_1/a$, the cell parameters being $a = 10.72$ Å; $b = 9.14$ Å; $c = 4.825$ Å; $\beta = 106°27'$; $\rho_{meas} = 1.537$ g/cm^3; $Z = 4$.

A satisfactory model was derived from the modified Patterson function (without the initial peak) by Patterson superposition. The structure was refined by least squares and from the three-dimensional electron-density distribution, the coordinates of the hydrogen atoms being deduced from crystallographic considerations. The final R(hk$l$) was 14.5%, the distances being correct to ~0.006 Å. Table 30 gives the atomic coordinates.

Figure 38 shows the distances and angles, which all agree well with those found for other amino acids. The C_2–N bond lies almost exactly in the plane of the carboxyl group, while the OH–group oxygen takes the position that brings it closest to both of the other functional groups.

The molecules are connected by framework of N–H...O and O–H...O bonds (Fig. 39). The bonds N–H...O(COO$^-$) and O(OH)–H...O(COO$^-$) connect the molecules into layers parallel to (100), while the N–H...O(OH) bonds cross-link the layers. The molecule takes a zwitterion form, because the hydrogen bonds around the nitrogen are tetrahedral, there are no O(COO$^-$)–H...O(COO$^-$) hydrogen bonds, and the carboxyl group has a symmetrical structure.

The structure of phosphoserine was determined in 1959 [69]. This is of interest in that it gives an indication of the character of the ester links to the serine, which play a major part in producing the tertiary structure of phosphoproteins.

The space group is $P2_12_12_1$, the cell parameters being $a = 7.79$ Å; $b = 10.24$ Å; $c = 9.09$ Å; $Z = 4$.

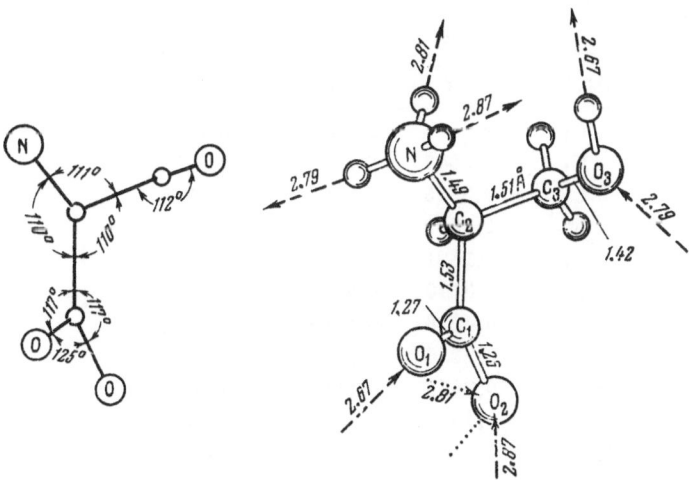

Fig. 38. Structure of the molecule in DL-serine.

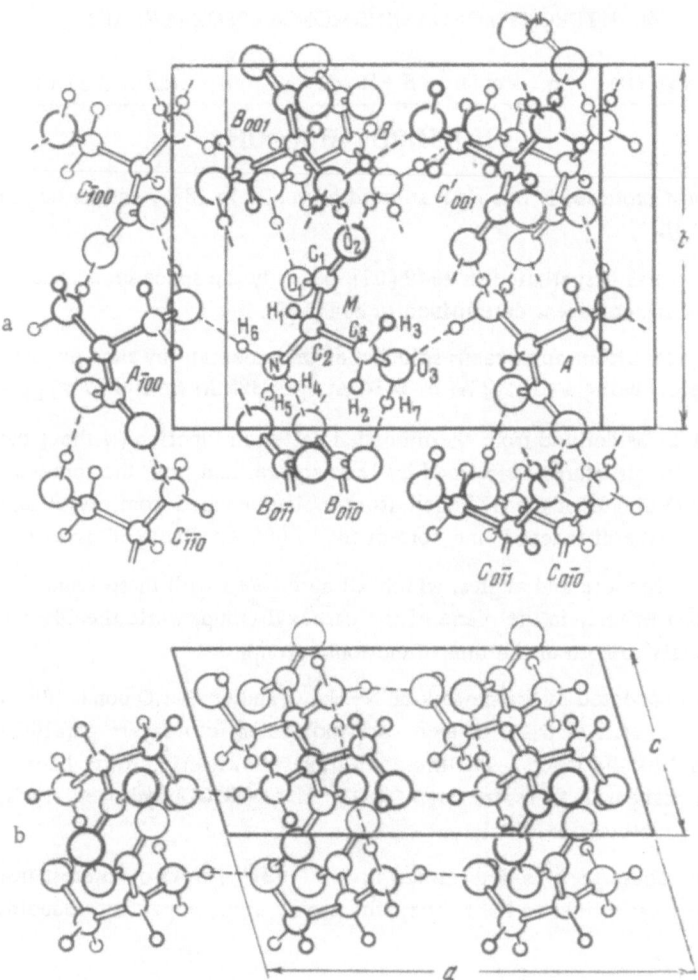

Fig. 39. Schematic representation of the structure of DL-serine;
(a) view perpendicular to xy plane; (b) view along b axis.

TABLE 30. Atomic Coordinates in the Structure of DL-Serine

Atom	x	y	z
C_1	0.2514	0.4047	0.1704
C_2	0.2547	0.2789	0.3808
C_3	0.3885	0.2100	0.4715
N	0.1532	0.1686	0.2460
O_1	0.1631	0.4021	0.0684
O_2	0.3340	0.5054	0.2533
O_3	0.4316	0.1683	0.2294
$H_1(C_2)$	0.2336	0.3222	0.5728
$H_2(C_3)$	0.3858	0.1147	0.6056
$H_3(C_3)$	0.4572	0.2889	0.5994
$H_4(N)$	0.1562	0.1182	0.0699
$H_5(N)$	0.1560	0.0988	0.4076
$H_6(N)$	0.0692	0.2232	0.2122
$H_7(O_3)$	0.3990	0.0708	0.1704

TABLE 31. Interatomic Distances in the
Structure of Phosphoserine

C_1-O_6	1.201 Å	C_3-O_1	1.466 Å
C_1-O_5	1.321	$P-O_1$	1.608
C_1-C_2	1.541	$P-O_2$	1.517
C_2-N	1.468	$P-O_3$	1.497
C_2-C_3	1.526	$P-O_4$	1.560

The structure model was derived by superposition from the Patterson function and was refined first via electron-density projections and then by three-dimensional least squares. The final R(hkl) was 10.3%. The interatomic distances are given in Table 31.

The bond lengths in the phosphate group imply resonance* between forms (I) and (II):

$$\begin{array}{cc}
O_3 & O_3^- \\
\| & | \\
HO_4-P-O_1 \quad (I) & HO_4-P-O_1 \quad (II) \\
| & \| \\
O_2 & O_2
\end{array}$$

This structure of the phosphate group, together with the asymmetry of the carboxyl group, implies a zwitterion form $^-HO_3POCH_2CH(NH_3)^+ \cdot COOH$. The configuration of the serine residue is analogous to that in DL-serine (carboxyl group, N, and C_2 coplanar, with the OH-group oxygen equidistant from the carboxyl and amino groups). Phosphoserine also has the planar group C_2, C_3, O_1, O_4, P, which lies almost perpendicular to the first group.

The packing (Fig. 40) involves a three-dimensional network of N—H...O bonds (lengths 2.79, 2.83, and 2.96 Å) and O—H...O bonds (2.47 and 2.55 Å).

Threonine (α-Amino-β-Hydroxybutyric Acid)

CH₃CH(OH)CH(NH₂)COOH

This is a biologically important amino acid found in most proteins; it was first detected by Zelinskii and Sadikov [18-20].

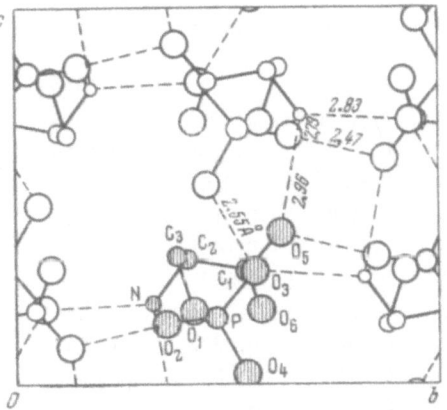

Fig. 40. Schematic representation of the
structure of phosphoserine; view along a axis.

---

*Assuming that a pure $P = O$ bond has a length of 1.469 Å, and P—OH one of 1.545 Å (from dibenzyl phosphate [69]).

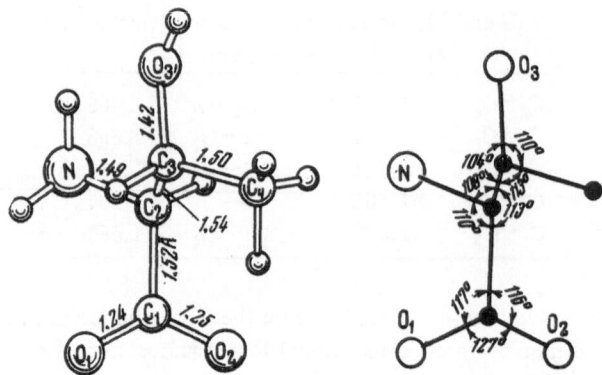

Fig. 41. Dimensions and structure of the threonine
molecule.

The molecule contains two asymmetric carbon atoms (α-C and β-C), so there are four stereoisomers; but only the stereoisomer shown in Fig. 41 is found in proteins.

The crystal structure was studied by Shoemaker et al. [70]. The crystals of L-threonine were grown by slow evaporation at constant temperature from aqueous solution. Space group $P2_12_12_1$, cell parameters a = 13.611 Å; b = 7.738 Å; c = 5.142 Å; ρ_{meas} = 1.464 g/cm^3; Z = 4. The model was derived from the three-dimensional Patterson function and was refined via the three-dimensional electron density and by least squares, the coordinates of the hydrogen atoms being deduced from crystallographic considerations, as well as from small peaks in the electron density. The final R(hkl) was 11.2%; distances to ~0.01 Å, angles to ~1°. Table 32 gives the coordinates.

Figure 41 shows the dimensions and structure of the threonine molecule.

TABLE 32. Atomic Coordinates in the Structure of L-Threonine

Atom	x	y	z
C_1	0.4956	0.1836	0.2984
C_2	0.3990	0.1074	0.3908
C_3	0.3178	0.2436	0.4310
C_4	0.2905	0.3346	0.1825
N	0.4131	0.0139	0.6415
O_1	0.5576	0.2168	0.4668
O_2	0.5026	0.2113	0.0590
O_3	0.2366	0.1468	0.5274
$H_1(C_2)$	0.387	0.007	0.260
$H_2(C_3)$	0.332	0.353	0.555
$H_3(C_4)$	0.230	0.427	0.235
$H_4(C_4)$	0.275	0.253	0.040
$H_5(C_4)$	0.363	0.380	0.090
$H_6(N)$	0.432	-0.100	0.585
$H_7(N)$	0.488	0.050	0.760
$H_8(N)$	0.340	-0.057	0.710
$H_9(O_3)$	0.155	0.266	0.535

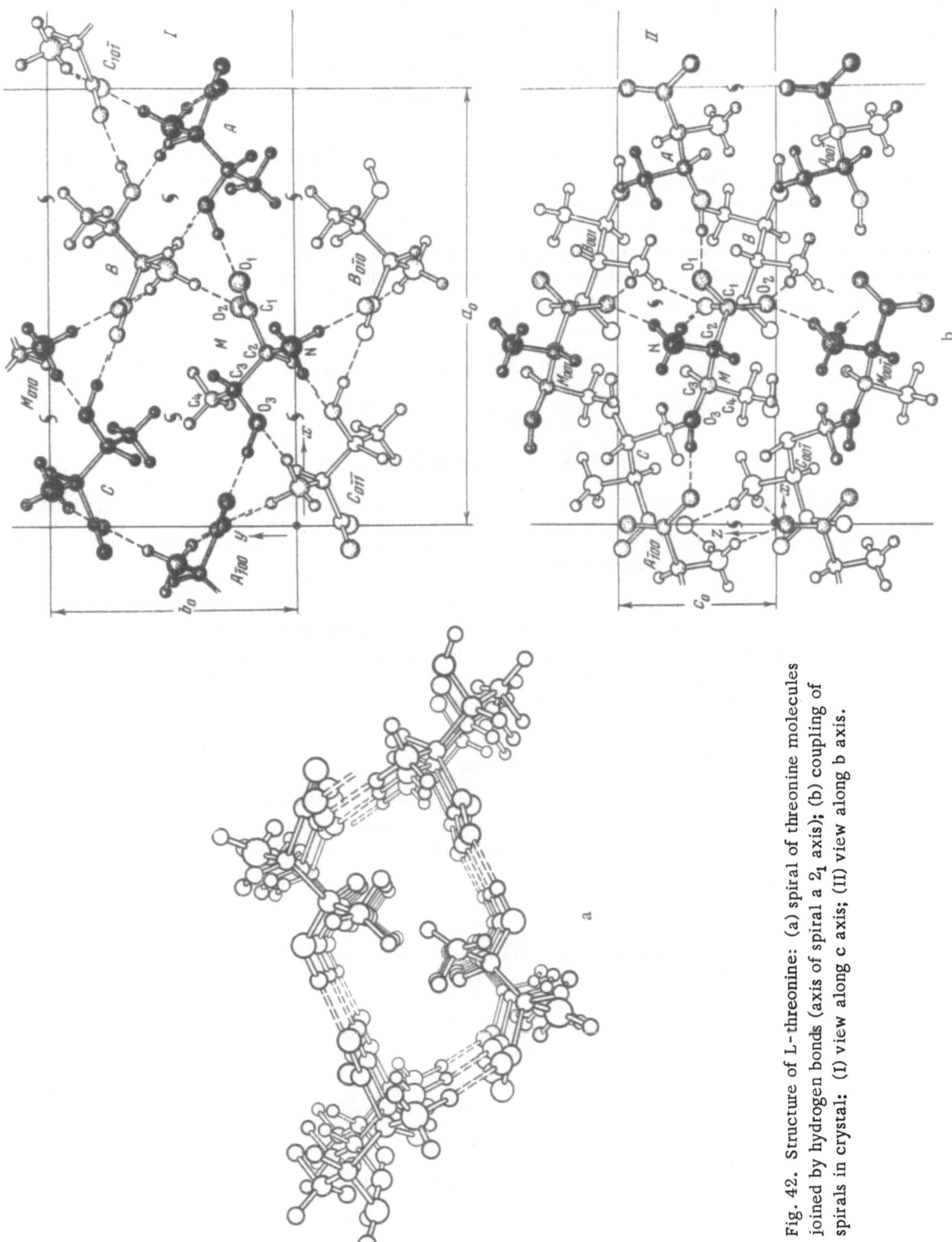

Fig. 42. Structure of L-threonine: (a) spiral of threonine molecules joined by hydrogen bonds (axis of spiral a 2_1 axis); (b) coupling of spirals in crystal: (I) view along c axis; (II) view along b axis.

The distances and angles are mainly close to the mean values found for other amino acids (see Tables 72-75), except for the short (1.50 Å) C_3-C_4 bond and the reduced (104°) $C_2-C_3-O_4$ angle.

The configuration is trans with respect to the C_2-C_3 bond; the hydrogen atom on C_3 lies between the NH_3 and COO^- groups, which is the most favorable position as regards steric hindrance. The carboxyl group is planar, the nitrogen atom lying 0.59 Å from the plane.

A spatial network of hydrogen bonds is present. The bonds $O_3-H...O_1'$ (2.66 Å) and $N-H...O_2$ (2.80 Å) join the molecules into infinite spirals parallel to the c axis (Fig. 42a). Within the spirals there are methyl groups in close contact (C_4-C_4' distances 3.79 Å), as in alanine. The $N-H...O_2''$ (2.90 Å) and $N-H...O_3'$ (3.10 Å) bonds link the spirals into a three-dimensional framework (Fig. 42b). There are also fairly short $N---O_1$ distances (3.08 Å), which do not correspond to ordinary hydrogen bonds, because they enclose no hydrogen atom, while the angles between these interatomic vectors and the valence bonds deviate greatly from tetrahedral. Here there may be a bifurcated hydrogen bond [25].

3. MONOAMINODICARBOXYLIC ACIDS

Aspartic Acid (α-Aminosuccinic Acid)

HOOCCH$_2$CH(NH$_2$)COOH

Aspartic acid occurs in many proteins; it was first isolated from conglutin by Ritthausen in 1868 [18-20].

Preliminary structural data were obtained in 1931 [23]; in 1951 the space group and cell parameters were established for the racemate and hydrochloride hemihydrate [61].

A more or less complete study was performed in 1959 for the racemate [71]: space group I2/a, cell parameters a = 9.18 Å; b = 7.49 Å; c = 15.79 Å; β = 96°; Z = 8. A satisfactory model was based on Patterson projections with extensive use of trial and error; some refinement was done via electron-density projections, but the considerable overlap caused the accuracy of the coordinates to be low [final values R(0kl) = 25%, R(h0l) = 20%] and we may say that only the general features are known. The preliminary coordinates of the main atoms are given in Table 33.

Figure 43 shows projections on the xz and yz planes.

The molecules extend along the c axis and are linked together by a complete system of $N-H...O$ and $O-H...O$ bonds, the $N-H...O$ bonds being between the amino group and the oxygen atoms of the carboxyl groups, with three bonds per nitrogen atom (one intramolecular, two intermolecular). The $O-H...O$ bonds are formed between the carboxyl groups of adjacent molecules.

TABLE 33. Coordinates of the Main Atoms in the Structure of DL-Aspartic Acid

Atom	x	y	z
C_1	0.038	0.161	0.049
C_2	0.079	0.094	0.134
C_3	0.000	-0.017	0.199
C_4	0.027	0.047	0.290
O_1	0.106	0.114	-0.011
O_2	-0.092	0.236	0.094
O_3	0.134	-0.060	0.308
O_4	-0.058	0.154	0.312
N	-0.133	0.081	0.208

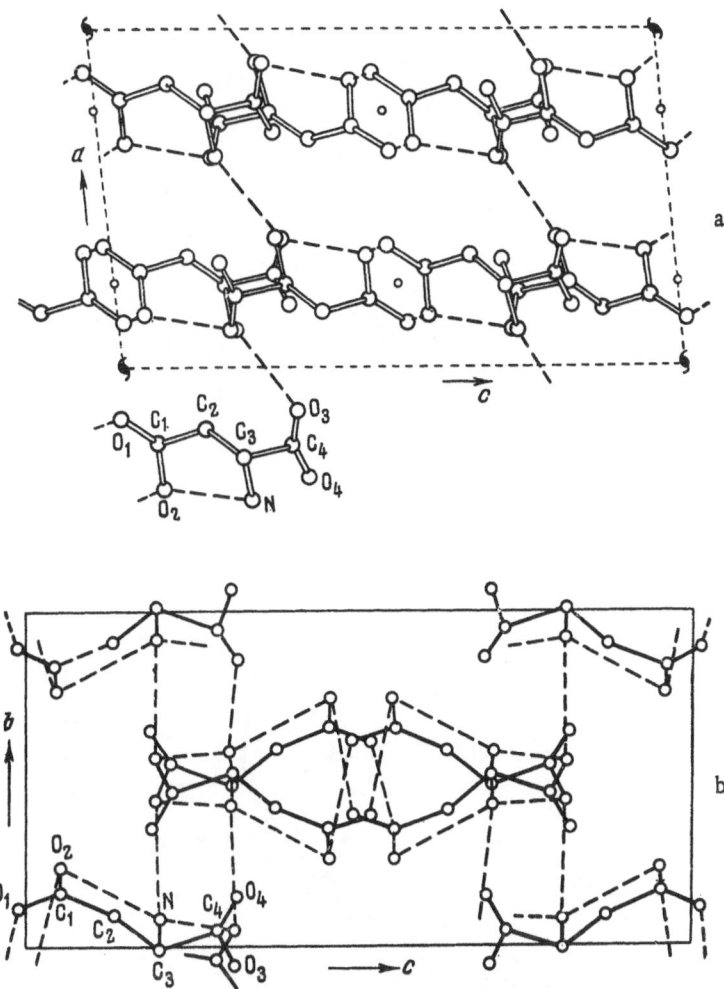

Fig. 43. Schematic representation of the structure of DL-aspartic
acid: (a) view along b axis, (b) view along a axis.

More precise data have been obtained from the isomorphous salts of L-aspartic acid with Zn^{2+}, Co^{2+}, and Ni^{2+} [72], which crystallize as trihydrates with space group $P2_12_12_1$ and the following cell parameters:

Zn · asp · 3H₂O	Co · asp · 3H₂O	Ni · asp · 3H₂O
a = 9.386Å	a = 9.39Å	a = 9.40Å
b = 7.920Å	b = 7.85Å	b = 7.83Å
c = 11.532Å	c = 11.37Å	c = 11.35Å
ρ_{meas} = 1.97 g/cm³	ρ_{meas} = 1.91 g/cm³	ρ_{meas} = 1.91 g/cm³
Z = 4	Z = 4	Z = 4

Doyne and Pepinsky deduced the absolute configuration for the cobalt salt on the basis of anomalous x-ray scattering. The resulting coordinates were assigned to the zinc salt and then were refined by least squares from the three-dimensional set of intensities. The final R(hkl) was 12%.

Figure 44 shows the distances and angles, as well as the absolute configuration of the L-aspartate ion. In this structure (as is usual in the salts of amino acids) each Zn^{2+} ion is surrounded by a distorted octahedron composed of nitrogen atoms and oxygen atoms from water molecules and COO⁻ groups. The acid residues are

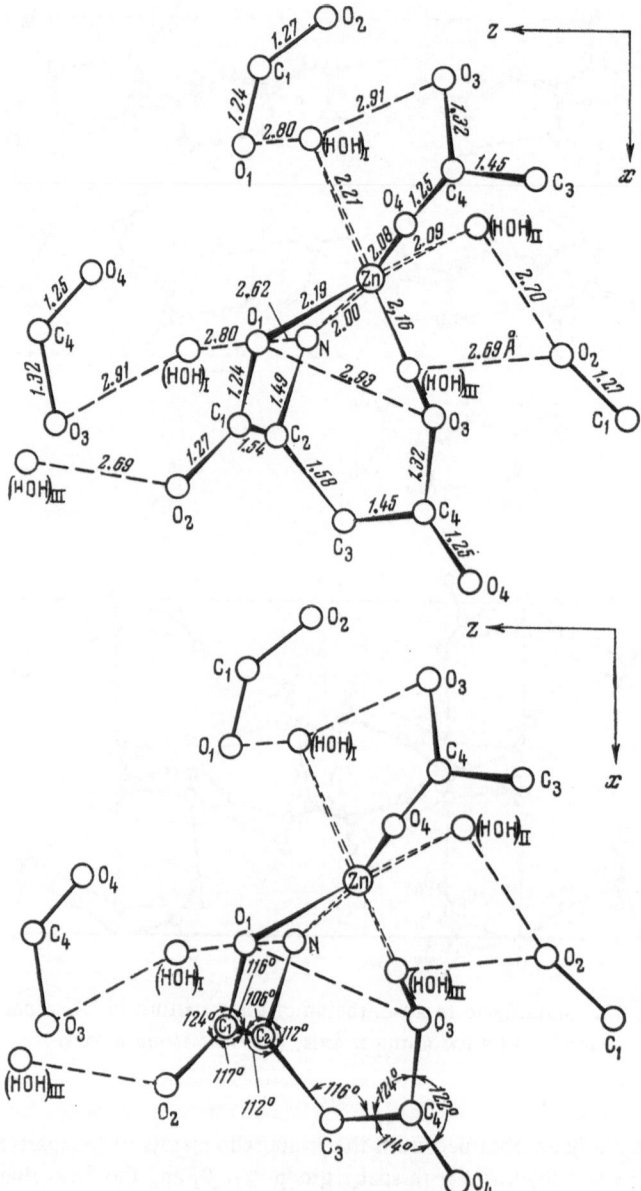

Fig. 44. Interatomic distances and valence angles in zinc
aspartate trihydrate.

convoluted, O_1, C_1, C_2, C_3, C_4, O_3, and H_1 forming a seven-member ring with an unusually short (1.45 Å) C_3–C_4 bond. These rings resemble the cyclic structures proposed by Steward and Thompson for asparagine, but which are not observed in its crystal structure.

Glutamic Acid (α-Aminoglutaric Acid)

HOOCCH$_2$CH$_2$CH(NH$_2$)COOH

This is one of the commoner amino acids; it was isolated from gliadin in 1866 by Ritthausen [18-20]. Dawson [73] examined its crystal structure in 1952. Crystals of the hydrochloride were grown by slow evapora-

TABLE 34. Coordinates of the Main Atoms in the Structure of Glutamic Acid Hydrochloride

Atom	x	y	z
C_1	0.379	0.144	0.410
C_2	0.562	0.195	0.334
C_3	0.477	0.157	0.227
C_4	0.669	0.183	0.145
C_5	0.580	0.126	0.047
O_1	0.422	0.036	0.428
O_2	0.209	0.198	0.451
O_3	0.425	0.051	0.044
O_4	0.701	0.168	-0.032
N	0.560	0.323	0.341
Cl^-	0.060	0.427	0.238

TABLE 35. Intramolecular Interatomic Distances and Valence Angles in the Structure of Glutamic Acid Hydrochloride

C_1-C_2	1.51 Å	$C_1-C_2-C_3$	109°
C_2-C_3	1.55	$C_2-C_3-C_4$	115
C_3-C_4	1.51	$C_3-C_4-C_5$	109
C_4-C_5	1.54	$C_2-C_1-O_2$	123
C_1-O_1	1.31	$C_2-C_1-O_1$	114
C_1-O_2	1.21	$O_1-C_1-O_2$	124
C_5-O_3	1.20	$C_4-C_5-O_3$	124
C_5-O_4	1.32	$C_4-C_5-O_4$	112
C_2-N	1.52	$O_3-C_5-O_4$	125
N atom 0.44 Å from		C_1-C_2-N	111
plane of $C_1O_1O_2$		C_3-C_2-N	110

tion from a solution of the racemate in 20% hydrochloric acid. Space group $P2_12_12_1$, cell parameters a = 5.16 Å; b = 11.8 Å; c = 13.30 Å; ρ_{meas} = 1.52 g/cm³; Z = 4. The absence of twofold symmetry elements and the presence of one formula unit in the independent part of the cell together indicate that the stereoisomers are separated during crystallization [74].

A trial model was deduced from Patterson projections, refinement being via successive electron-density projections. The final R were 10.8% for 0kl and 13.3% for h0l. The distances were determined to ~0.03 Å, the angles to ~3°. Table 34 gives the coordinates of the main atoms.

The distances and angles in the molecule are given in Table 35 (see Fig. 45 for notation). The carboxyl groups are clearly unsymmetrical with respect to the C_1-C_2 and C_4-C_5 bonds; also, the N atom lies out of the $C_1O_1O_2$ plane, and there is a systematic alternation of bond lengths* along the chain. This structure in the carboxyl groups indicates that the molecule takes the form [HOOCCH₂CH₂CH(NH₃)⁺COOH]. A framework of hydrogen bonds controls the packing in the crystal (Fig. 45). There are strong O−H...O bonds (2.57 A) between carboxyl groups, which join the molecules into zigzag chains parallel to the c axis. The bonds N−H...Cl⁻ (3.18 Å), N−H...O' (2.89 Å), and O−H...Cl⁻ (3.06 Å) cross-link the separate chains.

Hirokawa [75] determined the structure of L-glutamic acid in 1955. Space group $P2_12_12_1$, cell parameters a = 5.17 Å; b = 17.34 Å; c = 6.95 Å; ρ_{meas} = 1.57 g/cm³; Z = 4; ρ_x = 1.56 g/cm³. This was termed the β form, to distinguish it from the α form observed by Bernal [23], whose space group is $P2_12_12_1$ and cell parameters are a = 7.07 Å; b = 10.32 Å; c = 8.77 Å; Z = 4.

The model was deduced by minimizing Patterson projections together with Harker-Kasper sign inequalities, with refinement by electron-density projection and via differential syntheses, the coordinates of the hydrogen atoms being deduced from crystallographic considerations, and also via the difference series.

The minimum R were 16% for 0kl, 15.3% for h0l, and 15.4% for hk0. The distances were determined to ~0.05 Å and the angles to ~5°. Table 36 gives the atomic coordinates.

Figure 46 shows the structure of the molecule; the distances and angles fall within the limits usual for amino acids. The structure of the carboxyl groups corresponds to the zwitterion form HOOCCH₂CH₂CH(NH₃)⁺COO⁻. The general configuration is somewhat different from that found in·the hydrochloride, where the hy-

---

*This behavior of the C−C bonds occurs in norleucine and methionine, but a three-dimensional refinement is needed before a final conclusion can be drawn.

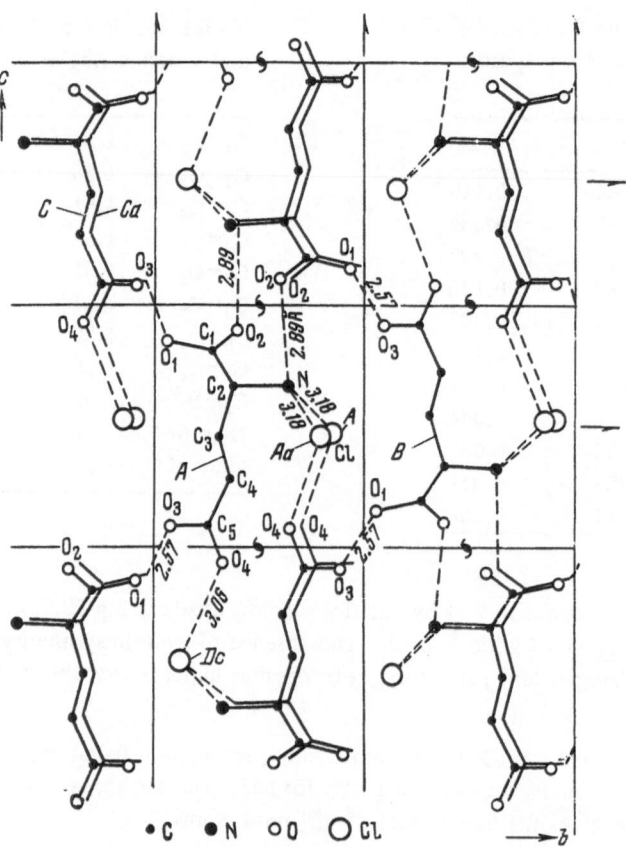

Fig. 45. Schematic representation of the structure of glutamic acid hydrochloride, view along a axis.

drocarbon chain has the trans configuration relative to the C_2-C_3 and C_3-C_4 bonds, whereas in the acid itself the configuration is trans with respect to C_2-C_3 and gauche with respect to C_3-C_4. The $C-N$ bond lies at 18° to the plane of O_1, C_1, O_2 in the hydrochloride but at 43° in the acid. There are roughly equal angles between the planes passing through the γ-carboxyl and $C_3C_4C_5$.

The molecules are bound together by $O-H...O$ and $N-H...O$ hydrogen bonds (Fig. 47).

TABLE 36. Atomic Coordinates in the Structure of L-Glutamic Acid

Atom	x	y	z	Atom	x	y	z
O_1	0.092	0.016	0.820	H_1	0.808	0.058	0.967
O_2	0.475	-0.045	0.842	H_2	0.757	0.128	0.085
C_1	0.325	0.013	0.863	H_3	0.570	0.046	0.138
C_2	0.463	0.091	0.899	H_4	0.667	0.373	0.677
C_3	0.571	0.123	0.706	H_5	0.323	0.130	0.961
C_4	0.750	0.194	0.719	H_6	0.667	0.075	0.628
C_5	0.607	0.273	0.754	H_7	0.411	0.133	0.613
O_3	0.746	0.325	0.692	H_8	0.888	0.184	0.832
O_4	0.400	0.276	0.830	H_9	0.862	0.197	0.590
N	0.663	0.079	0.043				

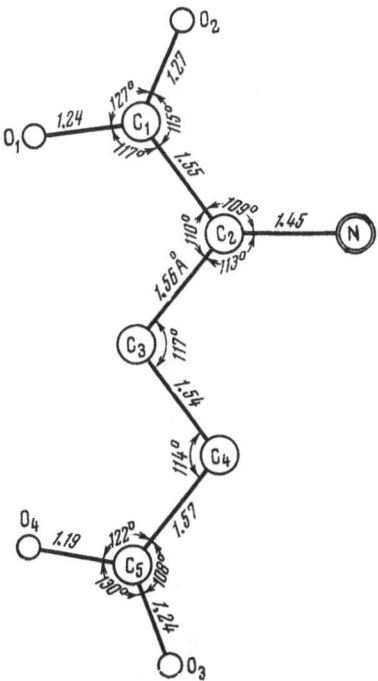

Fig. 46. Molecular configuration in
L-glutamic acid.

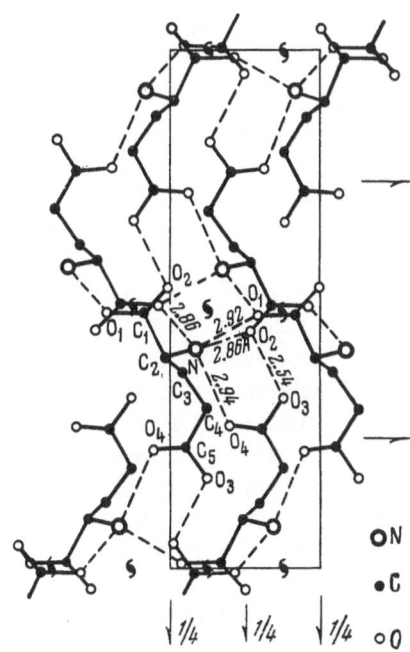

Fig. 47. Structure of L-glutamic acid
seen along c axis.

$O_3-H...O_2$ bonds (2.54 Å) join the molecules into chains parallel to the b axis, while $N-H...O$ bonds (2.86, 2.94, and 2.92 Å) connect the chains into a framework. There are also close $N---O$ contacts (2.86 Å), which arise by electrostatic interaction of NH_3^+ and COO^- groups in different molecules.

The structure of the glutamic acid residue has also been determined on copper and zinc glutamate dihydrates [76, 77].

Crystals of the copper salt were grown from aqueous solution by slow evaporation; space group $P2_12_12_1$, cell parameters a = 11.084 Å, b = 10.350 Å, c = 7.238 Å; ρ_{meas} = 1.954 g/cm^3; Z = 4 (Cu·glu·2H$_2$O); ρ_x = 1.957 g/cm^3.

The model was deduced from projections of the Patterson function, while the coordinates of the hydrogen atoms were deduced from a differential Fourier synthesis. The structure was refined by three-dimensional least squares, in the later stages based on 161 parameters, including the coordinates of the 24 atoms, anisotropic temperature factors for the 13 heavy atoms (Cu, O, N, C), and isotropic temperature factors for the hydrogen atoms (Table 37).

The final R(hk$l$) was 3.2%. The lengths of the Cu−O, Cu−N, C−O, C−N, and C−C bonds were determined to ~0.008 Å, while C−H, N−H, and O−H were determined to ~0.08 Å.

Figure 48 and Table 38 give the coordination of Cu^{2+} and the structure of the glutamate residue. The Cu^{2+} ion is surrounded by a very much distorted octahedron composed of the nitrogen and oxygen atoms of the water and the glutamate residue. Atoms O_1, O_3, O_5, and N are almost coplanar, the mean deviation from a plane being 0.025 Å. The Cu^{2+} ion lies 0.15 Å toward O_2 from the plane of the addend atoms; O_2 lies on the axis of the octahedron, while the opposite atom O_4 deviates strongly from this.

There are no special stereochemical features in the structure of the glutamate residue. Both carboxyl groups are planar within the error of measurement, the nitrogen atom lying 0.06 Å from the plane of the

TABLE 37. Atomic Parameters in the Structure of Copper Glutamate Dihydrate [Temperature Factors of the Form $T_j = \exp-(B_{11}h^2 + B_{22}k^2 + B_{33}l^2 + B_{12}hk + B_{13}hl + B_{23}kl)$]

Atom	x	y	z	$B_{11}\cdot10^4$	$B_{22}\cdot10^4$	$B_{33}\cdot10^4$	$B_{12}\cdot10^4$	$B_{13}\cdot10^4$	$B_{23}\cdot10^4$
Cu	0.7793	0.3167	0.6497	32	46	79	0	-11	-3
C_1	0.6595	0.5340	0.7780	43	39	71	-15	-15	21
C_2	0.5611	0.4661	0.6699	33	48	70	3	0	3
C_3	0.4630	0.4176	0.8015	31	58	95	3	10	20
C_4	0.5067	0.3326	0.9573	34	66	106	-2	0	48
C_5	0.4046	0.2829	1.0756	35	50	97	-5	12	-1
N	0.6133	0.3583	0.5602	38	57	78	-5	-3	-22
O_1	0.7666	0.4872	0.7683	36	58	113	-9	-20	-35
O_2	0.6310	0.6286	0.8741	54	46	112	0	-29	-45
O_3	0.4341	0.2214	1.2249	33	64·	98	5	13	49
O_4	0.2969	0.3006	1.0356	31	105	109	-5	9	57
$O_5(H_2O)$	0.7691	0.1344	0.5632	87	46	113	-8	-48	20
$O_6(H_2O)$	0.6559	0.0959	0.2343	42	65	113	10	-6	21

[Temperature Factors of the Form $T_j = \exp-(B_j\sin^2\theta/\lambda^2)$]

Atom	x	y	z	B_j	Atom	x	y	z	B_j
$H_1(C_2)$	0.528	0.522	0.583	-0.7	$H_7(N)$	0.554	0.282	0.574	2.5
$H_2(C_3)$	0.419	0.481	0.839	4.1	$H_8(O_5)$	0.781	0.084	0.610	2.2
$H_3(C_3)$	0.397	0.362	0.741	3.9	$H_9(O_5)$	0.716	0.125	0.431	5.9
$H_4(C_4)$	0.563	0.393	1.039	1.2	$H_{10}(O_6)$	0.699	0.123	0.197	6.2
$H_5(C_5)$	0.546	0.241	0.908	2.5	$H_{11}(O_6)$	0.581	0.129	0.227	1.1
$H_6(N)$	0.635	0.361	0.476	6.1					

TABLE 38. Lengths of Bonds Involving Hydrogen Atoms in the Structure of Copper Glutamate Dihydrate

C_2-H_1	0.93 Å	$N-H_6$	0.65 Å
C_3-H_2	0.86	$N-H_7$	1.03
C_3-H_3	1.03	O_5-H_8	0.63
C_4-H_4	1.06	O_5-H_9	1.13
C_4-H_5	1.10	O_6-H_{10}	0.61
		O_6-H_{11}	0.90

α-carboxyl. The $C_2-C_3-C_4$ and $C_3-C_4-C_5$ angles are somewhat larger than the tetrahedral angle, as is usual in amino acids. The side chain as a whole has an elongated configuration.

The units are linked together by a system of hydrogen bonds of medium strength (Fig. 49). A distinctive feature is that five of the six possible protons participate in the hydrogen bonds: four from the water molecules and one from the amino group, the second amino-group hydrogen atom, H_7, being free.

Crystals of zinc glutamate dihydrate were grown by slow evaporation from a saturated aqueous solution of zinc oxide in glutamic acid; space group $P2_12_12_1$, cell parameters a = 11,084 Å, b = 10.350 Å, c = 7.238 Å, Z = 4 (Zn·glu·$2H_2O$).

These cell parameters are very close to those for the copper compound, but the general intensity distributions show that the two structures are not identical. All the same, the initial stage of refinement by Fourier

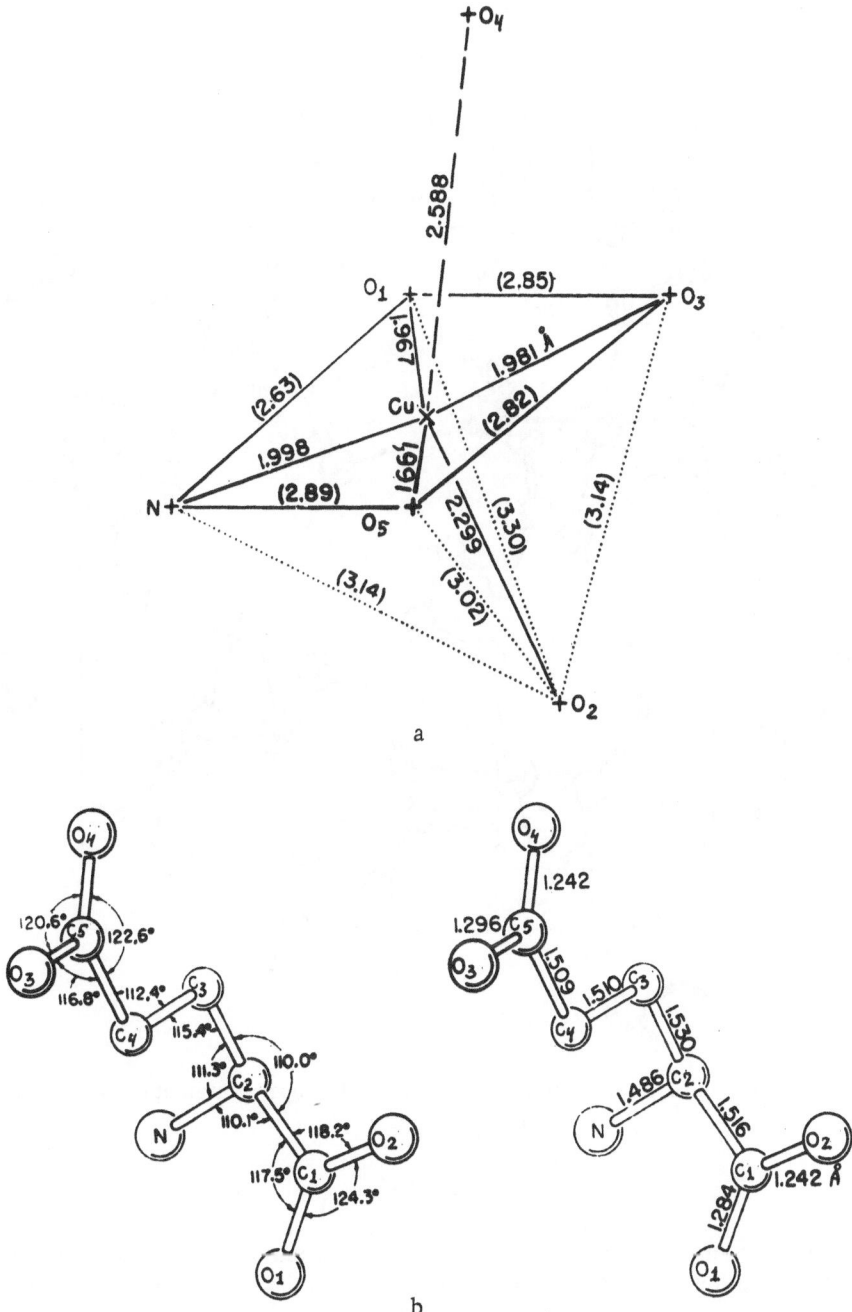

Fig. 48. Structure of copper glutamate dihydrate: (a) coordination of Cu^{2+}; (b) bond lengths and valence angles in the glutamate residue.

methods was based on the parameters of the copper salt as initial values. The final refinement was done by three-dimensional least squares, which gave the coordinates of the 24 atoms, the anisotropic temperature factors of the 13 heavy atoms (Zn, O, N, C), and the isotropic temperature factors of the 11 hydrogen atoms (Table 39). The final R(hk$l$) was 3.2%. The Zn—N, Zn—O, C—O, C—N, and C—C bond lengths were determined to ~0.008 Å, while the valence angles were determined to 0.4°.

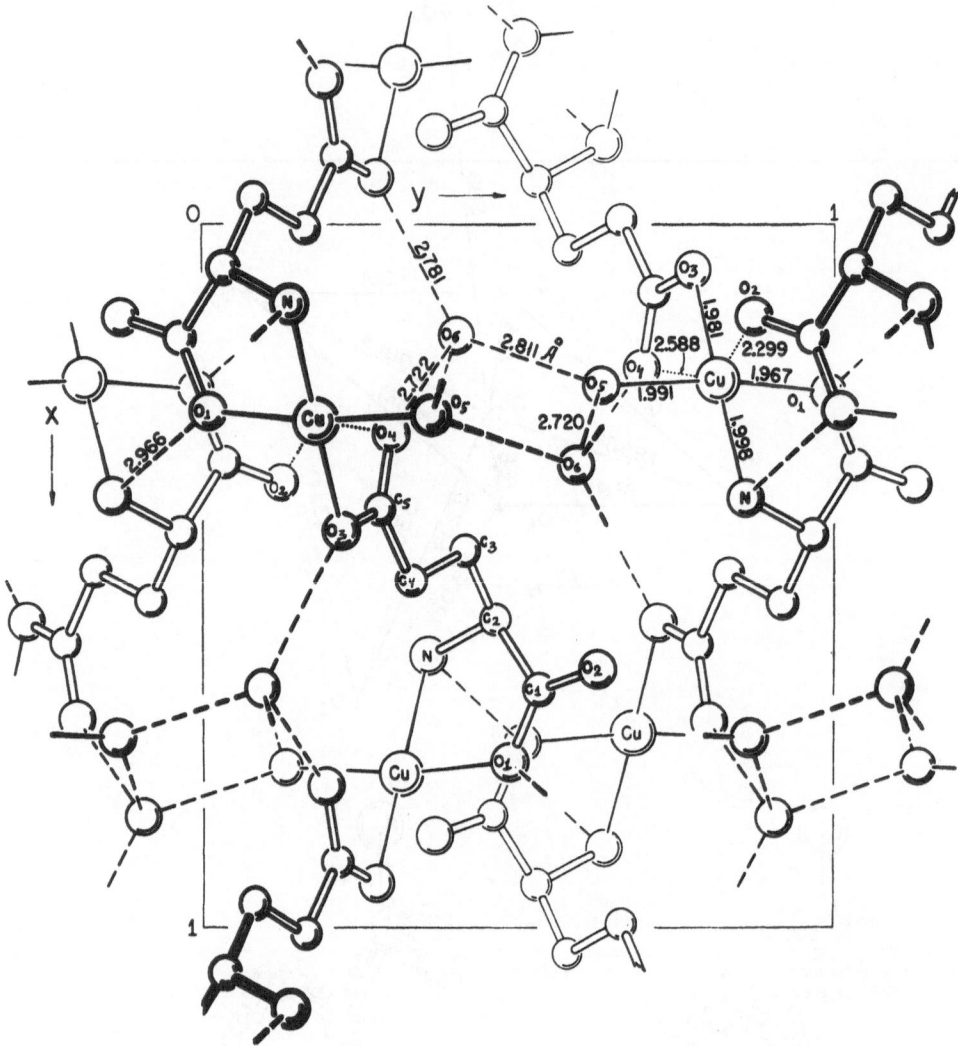

Fig. 49. Schematic representation of the structure of copper glutamate dihydrate seen
along c axis.

The main difference from the copper salt is that the coordination of the zinc ion is somewhat different (Fig. 50a). The Cu^{2+} ion lies only 0.15 Å from the plane of N, O_1, O_3, O_5, whereas Zn^{2+} lies 0.32 Å from that plane, which makes the Zn^{2+} equidistant (~ 2.069 Å) from five of the atoms of the coordination octahedron instead of four. The sixth atom, O_4, is 2.576 Å from the Zn^{2+} and deviates considerably from the axis of a regular octahedron. The bond lengths and valence angles in the glutamate residue (Fig. 50b and Table 40) agree within the limits of error with those for the copper salt, though there is some change in the size of the $C_1O_1O_2$ group, which may be ascribed to the change in the coordination of the metal. The shortening of the $Zn-O_2$ distance, i.e., the strengthening of the $Zn-O_2$ bond, is equivalent to a reduction in the order of the C_1-O_2 bond, which thus is 0.03 Å longer than that found for the copper salt. The order of the C_1-O_1 bond is correspondingly increased, with a shortening of 0.04 Å. These changes in the character of the $C-O$ bonds are reflected in the valence angles. In this case the nitrogen atom lies in the plane of the α-carboxyl.

The mode of packing in the crystal resembles that for the copper salt; only one hydrogen atom of the amino group is involved in forming a hydrogen bond, the second remaining free (Table 40).

TABLE 39. Atomic Parameters in the Structure of Zinc Glutamate Dihydrate [Temperature Factors of the Form $T_j = \exp-(B_{11}h^2 + B_{22}k^2 + B_{33}l^2 + B_{12}hk + B_{13}hl + B_{23}kl)$]

Atom	x	y	z	$B_{11} \cdot 10^4$	$B_{22} \cdot 10^4$	$B_{33} \cdot 10^4$	$B_{12} \cdot 10^4$	$B_{13} \cdot 10^4$	$B_{23} \cdot 10^4$
Zn	0.7908	0.3212	0.6279	38	49	91	6	-10	-4
C_1	0.6601	0.5303	0.7898	47	42	81	-14	-4	4
C_2	0.5605	0.4626	0.6814	34	58	82	1	14	-1
C_3	0.4641	0.4157	0.8128	32	68	106	-1	12	10
C_4	0.5096	0.3275	0.9639	41	84	123	-2	13	49
C_5	0.4110	0.2752	1.0879	39	59	123	-14	29	22
N	0.6102	0.3581	0.5674	35	67	99	-4	0	-28
O_1	0.7654	0.4935	0.7736	38	55	133	-1	-19	-53
O_2	0.6254	0.6212	0.8929	50	52	96	9	-15	-22
O_3	0.4458	0.2153	1.2342	43	81	112	-3	12	47
O_4	0.3028	0.2914	1.0554	40	94	139	-11	11	49
$O_5(H_2O)$	0.7697	0.1318	0.5517	97	49	134	-7	-68	22
$O_6(H_2O)$	0.6632	0.0976	0.2153	52	69	143	21	3	29

[Temperature Factors of the Form $T_j = \exp-(B_j \sin^2 \theta / \lambda^2)$]

Atom	x	y	z	B_j	Atom	x	y	z	B_j
$H_1(C_2)$	0.524	0.540	0.599	2.4	$H_7(N)$	0.574	0.291	0.602	1.5
$H_2(C_3)$	0.426	0.497	0.871	0.6	$H_8(O_5)$	0.795	0.084	0.586	0.1
$H_3(C_3)$	0.401	0.371	0.752	1.4	$H_9(O_5)$	0.755	0.130	0.456	5.4
$H_4(C_4)$	0.573	0.368	1.050	0.8	$H_{10}(O_6)$	0.701	0.130	0.167	2.2
$H_5(C_4)$	0.544	0.243	0.880	0.7	$H_{11}(O_6)$	0.607	0.174	0.229	6.7
$H_6(N)$	0.621	0.368	0.451	1.1					

TABLE 40. Lengths of Bonds Involving Hydrogen Atoms in the Structure of Zinc Glutamate Dihydrate

C_2-H_1	1.09 Å	$N-H_6$	0.86 Å	$N-H...O_1$	2.975 Å
C_3-H_2	1.04	$N-H_7$	0.85	$O_5-H...O_6$	2.729
C_3-H_3	0.95	O_5-H_8	0.62	$O_5-H...O_6'$	2.779
C_4-H_4	1.03	O_5-H_9	0.71	$O_6-H...O_3$	2.730
C_4-H_5	1.14	O_6-H_{10}	0.64	$O_6-H...O_4$	2.758
		O_6-H_{11}	1.02		

4. AMIDES OF MONOAMINODICARBOXYLIC ACIDS

Asparagine (β-Amide of α-Aminosuccinic Acid)

$H_2NC(O)CH_2CH(NH_2)COOH$

This was the first amino acid isolated from natural products (1806), but its presence in proteins was not demonstrated until 1932. The formula is very similar to that of glutamine (the difference being a single CH_2 group), but the chemical features are distinctive (occurrence in crystals only as hydrates, brown color with ninhydrin) [18-20]. It has several times been supposed that this anomalous behavior arises from a cyclic

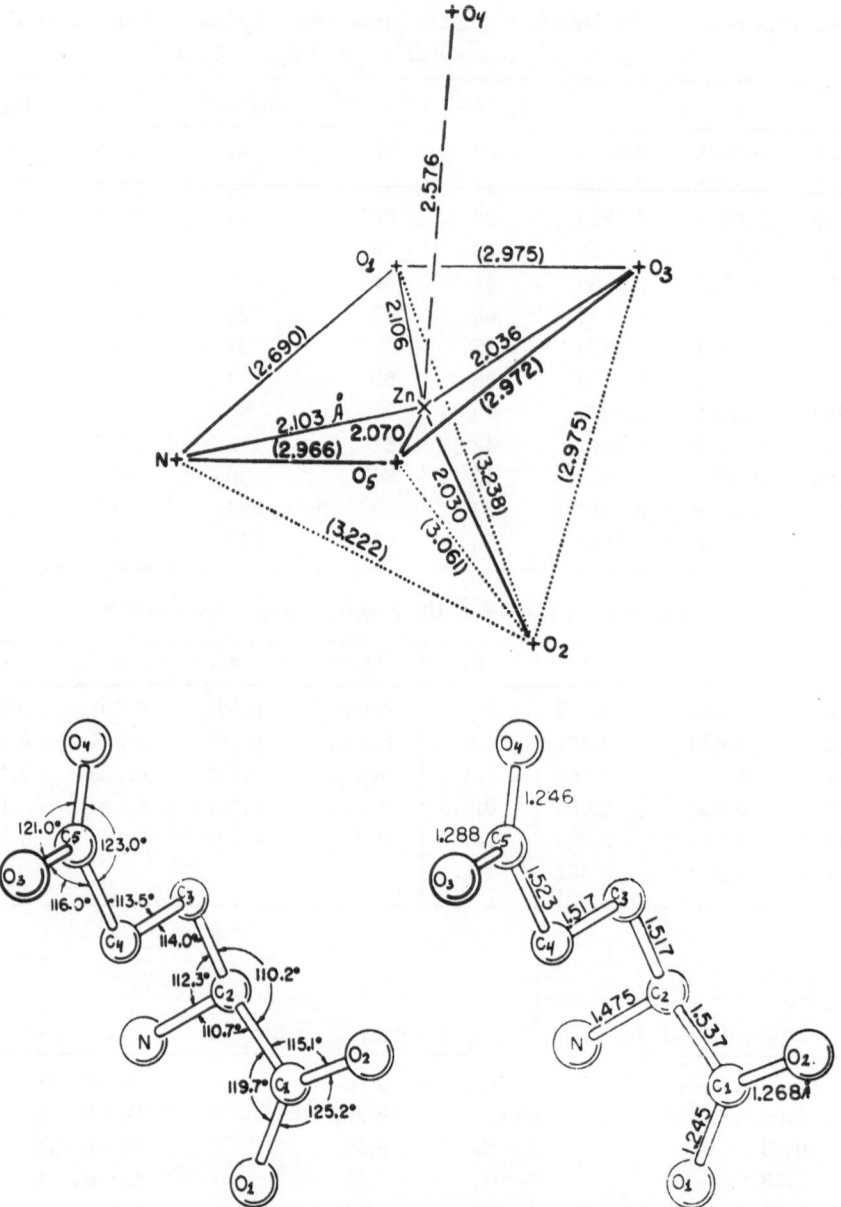

Fig. 50. Structure of zinc glutamate dihydrate: (a) coordination of Zn^{2+}; (b) bond lengths and valence angles in the glutamate residue.

structure; Steward and Thompson [78] supposed that the compound exists as the hydrate of the cyclic amide, while others [79] have ascribed the behavior to intramolecular interaction between the carboxyl and amide groups. Chemical evidence failed to provide unambiguous confirmation or rejection of these theories, so resort was made to x-ray methods.

The space group and unit cell were deduced in 1931 [23], and a preliminary structure study was performed in 1954 [79, 80], the structure being finally determined to fairly high accuracy in 1961 [81].

In every case the crystals were of asparagine monohydrate and had space group $P2_12_12_1$, the most accurate cell parameters [81] being $a = 5.582$ Å; $b = 9.812$ Å: $c = 11.796$ Å; $\rho_{meas} = 1.543$ g/cm^3; $Z = 4$ (asp · N · H$_2$O); $\rho_X = 1.543$ g/cm^3. A satisfactory model was derived from a sharpened three-dimensional Patterson diagram

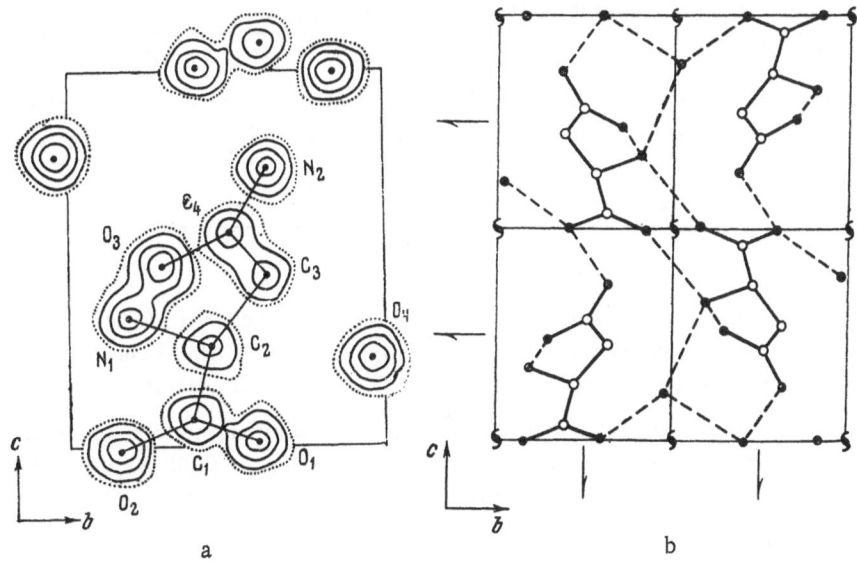

Fig. 51. Structure of L-asparagine monohydrate: (a) electron-density projection
on the xy plane; (b) view of structure along a axis.

TABLE 41. Coordinates of the Main Atoms in the Structure of Asparagine Monohydrate *

Atom	x	y	z
C_1	0.8757	0.9424	0.0357
C_2	0.7803	0.9731	0.1296
C_3	0.7889	0.0574	0.2271
C_4	0.0174	0.9982	0.2792
N_1	0.5834	0.8430	0.1720
N_2	0.1046	0.0641	0.3683
O_1	0.9949	0.0436	0.0007
O_2	0.8806	0.8246	0.9958
O_3	0.1112	0.8941	0.2389
O_4	0.3036	0.2301	0.1118

* The table of coordinates in [81] has the coordinates
incorrectly symbolized. The present symbols correspond to the cell axes.

and was refined via electron-density projections, differential syntheses, and three-dimensional least squares with anisotropic temperature factors. The final R(hk$l$) was 7.4%. Table 41 gives the coordinates of the main atoms.

Table 42 gives the principal interatomic distances, which are similar to analogous ones in other amino acids.

A distinctive feature is the elongated form of the molecule. The trans position of the carboxyl and amide groups relative to C_2-C_3 indicates that the Steward-Thompson cyclic structure is absent.

TABLE 42. Principal Interatomic Distances in the Structure of Asparagine Monohydrate

Intramolecular				Intermolecular			
C_1-O_1	1.27 Å	C_2-C_3	1.51 Å	$O_4-H...O_1$	2.84 Å	$N_1-H...O_4'$	2.85 Å
C_1-O_2	1.34	C_3-C_4	1.53	$O_4-H...O_1'$	2.80	$N_2-H...O_1'$	2.92
C_1-C_2	1.54	C_4-O_3	1.24	$N_1-H...O_3'$	2.80	$N_2-H...O_2'$	3.02
C_2-N_1	1.50	C_4-N_2	1.33	$N_1-H...O_2'$	2.81		

The packing is determined by a complete network of N—H...O and O—H...O bonds (Fig. 51), the O—H...O bond being between O_4 of the water and O_1 in the carboxyl group. The N—H...O bonds are produced by both nitrogen atoms with the carboxyl groups and by the amide nitrogen with the water.

The hydrogen atoms were located on differential electron-density projections, which showed that there are no intramolecular hydrogen bonds, in spite of the short N_1-O_3 distance (3.09 Å). This serves to show that there is no cyclic structure, but it does not prove that such a structure does not occur in solution.

However, the distances within the carboxyl group indicate some asymmetry in the latter. Here we find that the oxygen with a higher proportion of double-bond character participates in the stronger hydrogen bonds, whereas we would expect to find the converse, as in alanine.

Figure 53 illustrates the packing in the crystal. Each molecule forms five hydrogen bonds: two of them [$N_1-H...O_1$ (B) (2.94 Å); $N_1-H...O_3$ (C) (2.91 Å)] are produced by the NH_2 group and three [$N_2-H...O_2$ (A) (2.85 Å); $N_2-H...O_2$ (C) (2.79 Å); $N_2-H...O_1$ (B) (2.91 Å)] by the NH_3^+ group.

The structure reveals no stereochemical features that might explain the tendency of the molecule to cyclize.

Glutamine (γ-Amide of α-Aminoglutaric Acid)

$H_2NC(O)CH_2CH_2CH(NH_2)COOH$

This occurs in certain proteins; it also occurs in the free state in plants and animals.

A distinctive feature is the highly labile amide group, which leads to easy cyclization [18-20]. The crystal structure was examined in 1952 [82] to elucidate the reason for this.

The crystals were grown from an aqueous solution of L-glutamine. Space group $P2_12_12_1$, cell parameters a = 16.01 Å; b = 7.76 Å; c = 5.10 Å; ρ_{meas} = 1.54 g/cm³; Z = 4; ρ_X = 1.52 g/cm³.

The model was derived mainly by trial and error, refinement being via differential projection, the co-ordinates of the hydrogen atoms being deduced from crystallographic considerations, and also from the differential projections. The final R were 10% for hk0 and 13.5% for h0L. The distances were determined to ~0.02 Å. Table 43 gives the atomic coordinates.

Figure 52 gives the distances and angles in the molecule. The C_1-O_1 and C_1-N_1 bonds have 50% double-bond character, so there is resonance between forms I and II:

$$
\begin{array}{cc}
\overset{O_1}{\underset{\displaystyle -C_1}{\big\|}} & \quad\quad \overset{O_1^-}{\underset{\displaystyle -C_1}{\big|}} \\
\quad N_1H_2 \quad (I) & \quad\quad N_1H_2^+ \quad (II)
\end{array}
$$

The distribution of the hydrogen ions correspond to $RCH(NH_3)^+COO^-$.

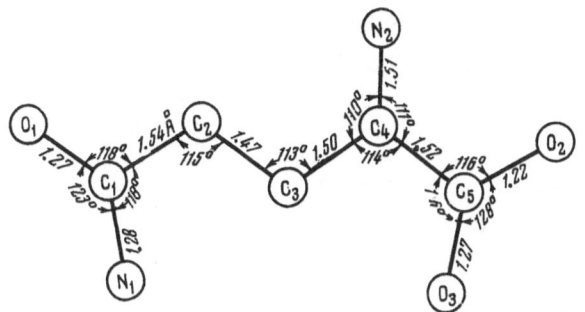

Fig. 52. Form of molecule in the structure of L-glutamine.

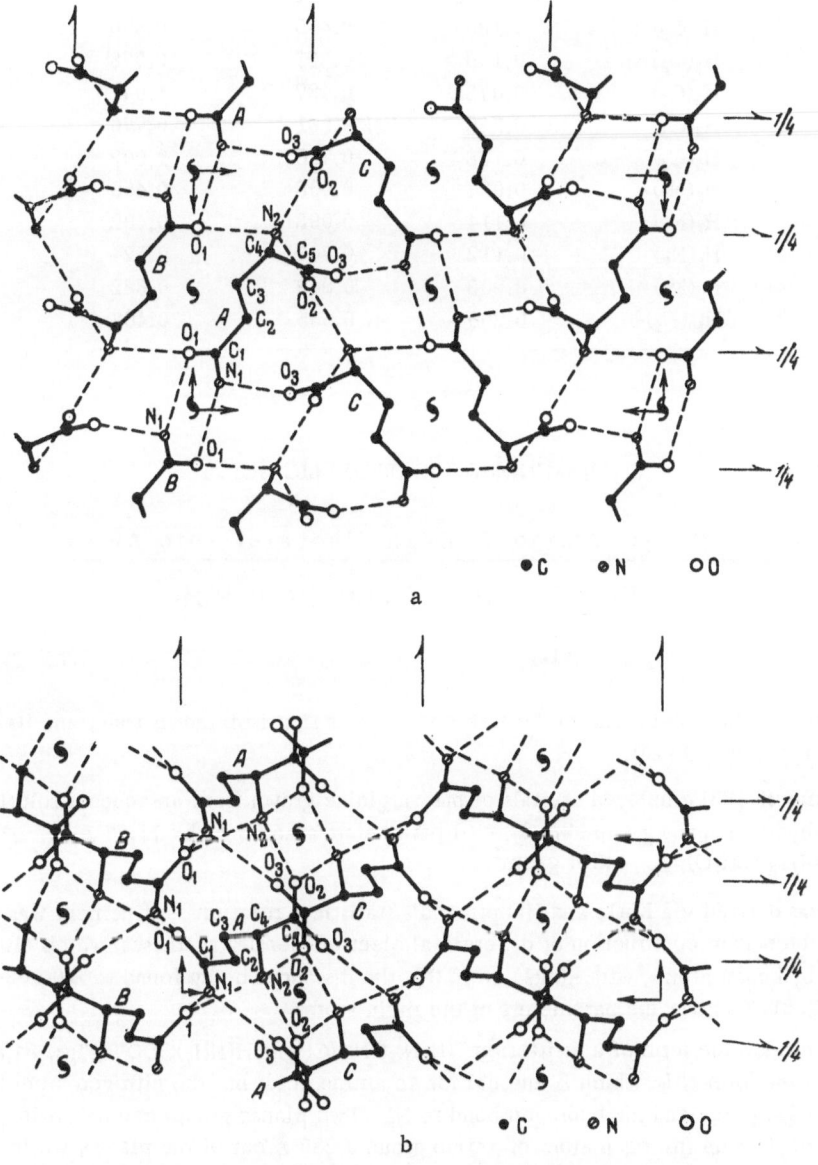

Fig. 53. Schematic representation of the structure of L-glutamine: (a) view
along c axis, (b) view along b axis.

TABLE 43. Atomic Coordinates in the Structure of
L-Glutamine

Atom	x	y	z
C_1	0.0481	0.2345	0.6960
C_2	0.1143	0.3775	0.7058
C_3	0.0912	0.5296	0.8607
C_4	0.1599	0.6610	0.8803
C_5	0.2417	0.5876	0.9823
N_1	0.0556	0.1173	0.5215
N_2	0.1724	0.7474	0.6176
O_1	-0.0087	0.2345	0.8686
O_2	0.2417	0.5399	1.2098
O_3	0.2986	0.5644	0.8117
$H_1(C_2)$	0.130	0.419	0.520
$H_2(C_2)$	0.169	0.331	0.778
$H_3(C_3)$	0.073	0.487	1.047
$H_4(C_3)$	0.036	0.581	0.776
$H_5(C_4)$	0.138	-0.756	1.002
$H_6(N_1)$	0.037	0.008	0.459
$H_7(N_1)$	0.114	0.093	0.402
$H_8(N_2)$	0.112	0.747	0.524
$H_9(N_2)$	0.203	0.849	0.682
$H_{10}(N_2)$	0.200	0.665	0.467

5. DIAMINOMONOCARBOXYLIC ACIDS

Arginine (α-Amino-δ-Guanidinovalerianic Acid)

$$HN = CNHCH_2CH_2CH_2CH\,(NH_2)COOH$$
$$\underset{NH_2}{\mid}$$

This occurs widely in proteins and in the free state; it was first isolated in 1866, and its chemical structure was established in 1897 [18-20].

The x-ray analysis [83] employed crystals of pure arginine grown from an aqueous solution of the L form. The product is a dihydrate; space group $P2_12_12_1$, cell parameters a = 5.68 Å; b = 11.87 Å; c = 15.74 Å; ρ_{meas} = 1.320 g/cm^3; Z = 4 (arg · 2H$_2$O); ρ_x = 1.314 g/cm^3.

The model was derived via Karle and Hauptmann's statistical relations; refinement was by least squares, with squares, with subsequent construction of differential electron-density syntheses, which also gave the coordinates of the hydrogen atoms, with R(hk$l$) of 10.3%, the distances being found to ~0.008-0.012 Å and the angles to ~0.9°. Table 44 gives the parameters of the main atoms.

The molecule takes the form of a zwitterion $^+$(H$_2$N)$_2$CNH(CH$_2$)$_3$CH(NH$_2$)COO$^-$ (Figs. 54, and 55), the amino group having the form NH$_2$, which is unusual for an amino acid; but the nitrogen atom has a tetrahedral environment, with C$_4$, H$_2$, H$_3$, and the hydrogen bond to N$_9^l$. Two planar groups can be distinguished: the first consists of the carboxyl group (nitrogen atom of amino group 0.280 Å out of the plane), while the second consists of the atoms in the carbon chain and all atoms in the guanidine group, including the hydrogen atoms. The planes meet at 74°.

TABLE 44. Parameters of the Main Atoms in the Structure of L-Arginine Dihydrate [Temperature Factors of the Form $T_j = \exp-(B_{11}h^2 + B_{22}k^2 + B_{33}l^3 + 2B_{12}hk + 2B_{13}hl + B_{23}kl)$]

Atom	x	y	z	$10^4 \cdot B_{11}$	$10^4 \cdot B_{22}$	$10^4 \cdot B_{33}$	$10^4 \cdot B_{12}$	$10^4 \cdot B_{13}$	$10^4 \cdot B_{23}$
O_1	0.3022	0.4036	0.3231	244	44	42	23	68	4
O_2	0.3340	0.5832	0.2830	219	45	31	6	53	2
C_3	0.2353	0.5047	0.3220	162	41	17	-2	16	3
C_4	0.0072	0.5293	0.3724	95	35	14	-4	6	-3
N_5	-0.0839	0.6432	0.3524	136	44	23	7	0	-2
C_6	0.0622	0.5186	0.4680	149	56	12	-31	1	1
C_7	0.2546	0.5995	0.5002	288	59	19	-45	-52	11
C_8	0.3822	0.5478	0.5753	148	52	22	-14	-39	11
N_9	0.5532	0.6274	0.6118	127	37	14	-3	-24	1
C_{10}	0.7315	0.5955	0.6633	110	46	10	-12	8	-2
N_{11}	0.7796	0.4862	0.6760	161	36	16	9	-6	6
N_{12}	0.8579	0.6739	0.7016	191	41	24	-1	-52	1
$O(H_2O)_1$	0.5506	0.2790	0.6214	252	45	66	17	25	6
$O(H_2O)_2$	0.6222	0.2871	0.4356	323	195	50	89	42	28

Coordinates of the Hydrogen Atoms

Atom	x	y	z	Atom	x	y	z
$H_1(C_4)$	0.908	0.458	0.350	$H_{10}(N_9)$	0.525	0.703	0.617
$H_2(N_5)$	0.847	0.650	0.328	$H_{11}(N_{11})$	0.722	0.433	0.660
$H_3(N_5)$	0.938	0.645	0.362	$H_{12}(N_{11})$	0.862	0.472	0.705
$H_4(C_6)$	0.076	0.442	0.475	$H_{13}(N_{12})$	0.933	0.663	0.719
$H_5(C_6)$	0.933	0.525	0.495	$H_{14}(N_{12})$	0.822	0.741	0.700
$H_6(C_7)$	0.366	0.645	0.450	$H_{15}(H_2O)_1$	0.533	0.207	0.629
$H_7(C_7)$	0.208	0.682	0.502	$H_{16}(H_2O)_1$	0.459	0.267	0.628
$H_8(C_8)$	0.388	0.478	0.589	$H_{17}(H_2O)_2$	0.558	0.335	0.482
$H_9(C_8)$	0.253	0.512	0.623	$H_{18}(H_2O)_2$	0.558	0.275	0.463

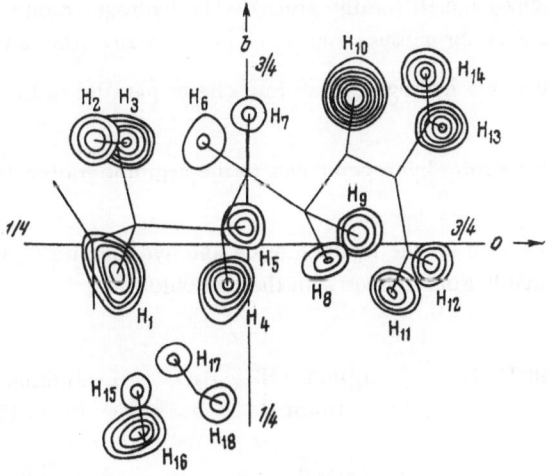

Fig. 54. Differential-synthesis projection designed to determine the positions of the hydrogen atoms in the structure of L-arginine dihydrate. The lines are at intervals of 0.05 e/A³, with the first at 0.15 e/A³.

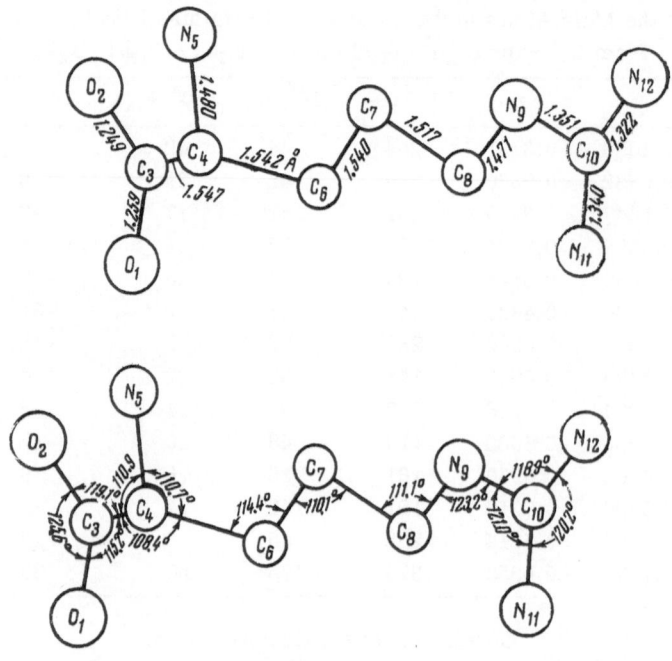

Fig. 55. Molecular shape in the structure of L-arginine
dihydrate.

The mean C—C bond length (1.537 Å) is very close to the standard value; the mean C—C—C angle is 111°, with C_4—C_6—C_7 (114.4°) showing the largest deviation. Some analogous increase in the angle adjacent to the amino group occurs in other amino acids.

The C_8—N_9 bond is 1.471 Å; the C—N bonds within the guanidine group are nearly equal (mean 1.338 Å). The bond lengths and valence angles in the carboxyl group are close to those for α-glycine and L-lysine · HCl · $2H_2O$.

Figure 56 shows the packing; the molecules extend along the c axis, being joined into chains via hydrogen bonds between the carboxyl group and the nitrogen of the guanidine group. These two groups are coplanar. Along the b axis the molecules are joined via hydrogen bonds between O_2 (carboxyl) and $N_{12}H_{14}$ (guanidine), as well as ones between N_9H_{10} (guanidine) and N_5 (amino group). The hydrogen atoms of the amino group do not participate in hydrogen bonds; this is rather unusual and is not found in any other amino acid.

The water molecules are linked via hydrogen bonds into chains parallel to the a axis, which are perpendicular to the arginine chains.

In addition, the water molecules form hydrogen bonds to the arginine molecules. The four hydrogen bonds around $O(H_2O)_2$ are coplanar.

Molecular structures are also known for the hydrochloride and hydrobromide [84-86]. The hydrochloride crystallize in two forms, one of which is isomorphous with the hydrobromide:

L-arginine · HBr · H_2O	L-arginine · HCl · H_2O (form I)	L-arginine · HCl (form II)
a = 11.26 Å	a = 11.22 Å	a = 5.33 Å
b = 8.65	b = 8.50	b = 9.46
c = 11.25	c = 11.07	c = 20.07
β = 91.5°	β = 91°	β = 90.5°
Z = 4 (arg · HBr · H_2O)	Z = 4 (arg · HCl · H_2O)	Z = 4 (arg · HCl)

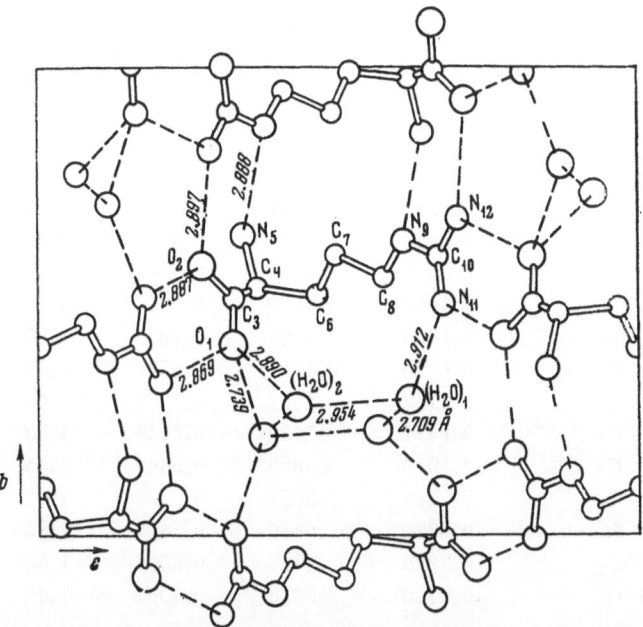

Fig. 56. The structure of L-arginine dihydrate seen along
the a axis.

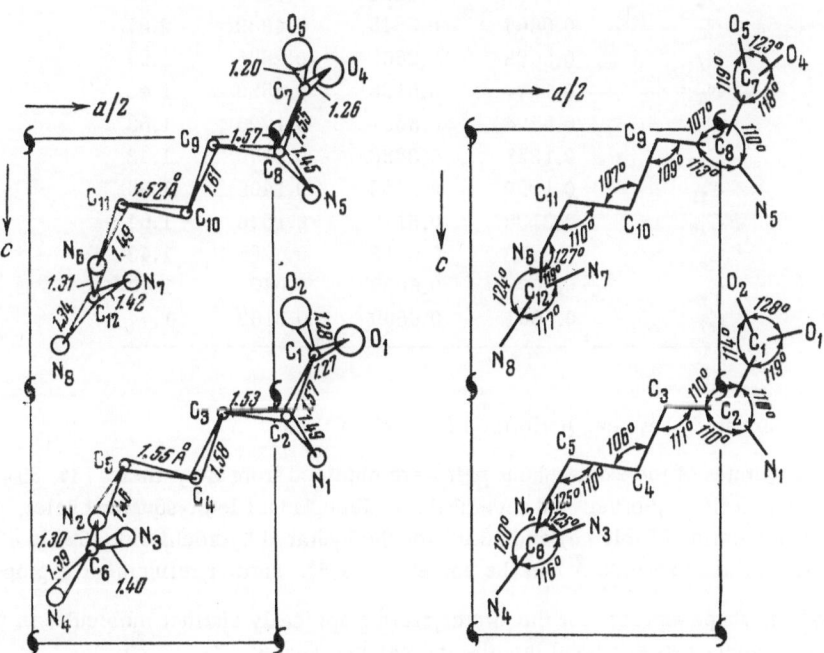

Fig. 57. Form of the crystallographically nonequivalent arginine molecules
in L-arginine · HBr · H₂O.

TABLE 45. Parameters of the Main Atoms in the Structure of L-Arginine Hydrobromide Monohydrate (Temperature Factors of the Form $T_j = \exp - B_j \sin^2 \theta / \lambda^2$)

Atom	x	y	z	B_j, $\overset{\circ}{A}{}^2$
		First molecule		
Br_1^-	0.1679	0.2474	0.8726	1.91
$O_3(H_2O)$	0.1198	0.1341	0.5890	2.83
O_2	0.5415	0.6654	0.3578	1.88
O_1	0.6491	0.4493	0.4095	2.09
N_4	0.0422	0.7509	0.9054	1.95
N_3	0.1826	0.8771	0.7878	2.80
N_2	0.1320	0.6117	0.7560	1.64
N_1	0.5758	0.4511	0.6357	1.36
C_6	0.1221	0.7405	0.8141	1.59
C_5	0.1943	0.5955	0.6441	1.55
C_4	0.3288	0.5681	0.6702	1.63
C_3	0.3876	0.5510	0.5452	1.89
C_2	0.5225	0.5719	0.5561	1.35
C_1	0.5789	0.5610	0.4302	1.45
		Second molecule		
Br_2^-	0.1284	0.8902	0.3623	1.53
O_6	0.1199	0.9724	0.0827	2.60
O_5	0.5458	0.7647	0.8506	2.45
O_4	0.6093	0.5324	0.8850	2.59
N_8	0.0607	0.3648	0.4089	2.67
N_7	0.2098	0.2665	0.2875	1.97
N_6	0.1371	0.5186	0.2580	1.43
N_5	0.5724	0.5601	0.1226	1.53
C_{12}	0.1321	0.3888	0.3170	1.43
C_{11}	0.1922	0.5435	0.1406	1.86
C_{10}	0.3138	0.6172	0.1576	1.64
C_9	0.3722	0.6213	0.0285	1.43
C_8	0.5072	0.6632	0.0424	1.40
C_7	0.5601	0.6569	0.9163	1.74

A complete structure study has been performed for each of these.

Models for the structures of the isomorphous pair were obtained from β-syntheses [84, 85], with refinement via differential-synthesis projections and several three-dimensional least-squares cycles. The R(hk$l$) were 11.4% for the hydrobromide (Table 45) and 13.9% for the hydrated hydrochloride (Table 46), the distances in both cases being determined to ~0.03 Å and the angles to ~1.8°. Further refinement is proposed.

Figure 57 shows the shape and size for the two crystallographically distinct molecules in the hydrobromide; there is good agreement between bond lengths and valence angles.

Both molecules lie parallel to the ac plane and extend diagonally in that plane. The main difference between the two lies in the orientation of the guanidine groups. In the first molecule this group (C_6, N_2, N_3, N_4) lies along the positive direction of the b axis, while in the second (C_{12}, N_6, N_7, N_8) it lies along the negative direction.

TABLE 46. Parameters of the Main Atoms in the Structure of L-Arginine Hydrochloride Monohydrate

First molecule				Second molecule			
Atom	x	y	z	Atom	x	y	z
Cl_1^-	0.1609	0.2434	0.8689	Cl_2^-	0.1312	0.8972	0.3618
$O_3(H_2O)$	0.1206	0.1433	0.5973	$O_6(H_2O)$	0.1253	0.9765	0.0902
O_2	0.5399	0.6718	0.3557	O_5	0.5362	0.7780	0.8438
O_1	0.6539	0.4576	0.4112	O_4	0.6148	0.5423	0.8865
N_4	0.0379	0.7500	0.9120	N_8	0.0579	0.3686	0.4138
N_3	0.1807	0.8699	0.7952	N_7	0.2061	0.2672	0.2936
N_2	0.1246	0.6079	0.7629	N_6	0.1348	0.5200	0.2631
N_1	0.5843	0.4575	0.6369	N_5	0.5782	0.5755	0.1250
C_6	0.1152	0.7436	0.8210	C_{12}	0.1318	0.3922	0.3183
C_5	0.1911	0.5908	0.6462	C_{11}	0.1898	0.5449	0.1430
C_4	0.3296	0.5747	0.6739	C_{10}	0.3174	0.6205	0.1596
C_3	0.3856	0.5561	0.5454	C_9	0.3706	0.6274	0.0305
C_2	0.5253	0.5815	0.5568	C_8	0.5059	0.6787	0.0408
C_1	0.5798	0.5702	0.4327	C_7	0.5564	0.6697	0.9150

Interatomic Distances and Valence Angles in the Structure of L-Arginine Hydrochloride Monohydrate

First molecule				Second molecule			
N_4-C_6	1.34 Å	$\angle N_4C_6N_3$	119°42'	N_8-C_{12}	1.37 Å	$\angle N_8C_{12}N_7$	128°42'
C_6-N_3	1.33	$N_3C_6N_2$	123 02	$C_{12}-N_7$	1.42	$N_7C_{12}N_6$	123 30
C_6-N_2	1.32	$N_4C_6N_2$	116 37	$C_{12}-N_6$	1.25	$N_8C_{12}N_6$	121 48
N_2-C_5	1.51	$C_6N_2C_5$	123 28	N_6-C_{11}	1.49	$C_{12}N_6C_{11}$	125 00
C_5-C_4	1.58	$N_2C_5C_4$	109 51	$C_{11}-C_{10}$	1.58	$N_6C_{11}C_{10}$	109 57
C_4-C_3	1.57	$C_5C_4C_3$	103 52	$C_{10}-C_9$	1.56	$C_{11}C_{10}C_9$	105 35
C_3-C_2	1.59	$C_4C_3C_2$	106 18	C_9-C_8	1.58	$C_{10}C_9C_8$	108 52
C_2-N_1	1.52	$C_3C_2N_1$	113 24	C_8-N_5	1.51	$C_9C_8N_5$	113 00
C_2-C_1	1.52	$N_1C_2C_1$	107 54	C_8-C_7	1.52	$N_5C_8C_7$	107 06
C_1-O_1	1.29	$C_3C_2C_1$	106 42	C_7-O_4	1.31	$C_9C_8C_7$	105 46
C_1-O_2	1.28	$C_2C_1O_2$	114 30	C_7-O_5	1.23	$C_8C_7O_5$	115 36
		$O_2C_1O_1$	126 48			$O_5C_7O_4$	123 28
		$O_1C_1C_2$	118 33			$O_4C_7C_8$	120 45

The hydrogen bonds between molecules are as follows: $O-H...Br^-$, $N-H...O$, $N-H...Br^-$, $O(H_2O)-H...O$.

The hydrochloride monohydrate of L-arginine has a structure analogous to that of the hydrobromide, so Table 46 gives the basic data for this without discussion.

The results of Table 46 for the hydrochloride monohydrate [86] have not been published; they have kindly been made available by workers at Madras University.

The distances and angles have been determined to ~0.015 Å and ~0.9°, respectively. The bond lengths (Table 47) agree well with those found for arginine dihydrate.

It is of interest to compare the configuration and the distribution of the hydrogen atoms for these structures.

All have the guanidine and carboxyl groups planar, but the relative disposition of these groups is affected by rotation around the $C_\alpha-C_\beta$ and $C_\delta-N_\epsilon$ bonds. Three configurations occur for the γ-carbon (two gauche,

TABLE 47. Parameters of the Main Atoms in the Structure of L-Arginine Hydrochloride

First molecule				Second molecule			
Atom	x	y	z	Atom	x	y	z
Cl_1^-	0.1159	0.0000	0.1743	Cl_2^-	0.1242	0.4996	0.3239
O_2	0.2960	0.3349	-0.0158	O_4	0.7950	0.6978	0.5405
O_1	0.3370	0.5551	0.0238	O_3	0.4075	0.7776	0.5209
N_4	0.3445	0.1145	0.3239	N_8	0.3208	0.6229	0.1687
N_3	0.6896	0.2616	0.3375	N_7	0.6662	0.7684	0.1562
N_2	0.5350	0.2030	0.2318	N_6	0.5219	0.7136	0.2610
N_1	0.8357	0.5455	-0.0185	N_5	0.9801	0.9673	0.5208
C_6	0.5273	0.1967	0.2988	C_{12}	0.5052	0.7063	0.1951
C_5	0.7130	0.2983	0.1976	C_{11}	0.7071	0.8015	0.2975
C_4	0.6404	0.2935	0.1229	C_{10}	0.6407	0.8037	0.3709
C_3	0.7940	0.4059	0.0842	C_9	0.8366	0.8805	0.4107
C_2	0.7098	0.4215	0.0112	C_8	0.7564	0.9183	0.4809
C_1	0.4240	0.4383	0.0038	C_7	0.6413	0.7854	0.5172

Interatomic Distances and Valence Angles in the Structure of L-Arginine Hydrochloride

First molecule				Second molecule			
C_6-N_4	1.348 Å	$\angle N_3C_6N_4$	121°36'	N_8-C_{12}	1.362 Å	$\angle N_8C_{12}N_7$	120°18'
C_6-N_3	1.311	$N_2C_6N_4$	115 18	$C_{12}-N_7$	1.306	$N_8C_{12}N_6$	117 24
C_6-N_2	1.348	$N_2C_6N_3$	123 00	$C_{12}-N_6$	1.328	$N_7C_{12}N_6$	112 12
C_5-N_2	1.483	$C_6N_2C_5$	121 00	N_6-C_{11}	1.479	$C_{12}N_6C_{11}$	124 12
C_5-C_4	1.543	$N_2C_5C_4$	106 00	$C_{11}-C_{10}$	1.516	$N_6C_{11}C_{10}$	109 12
C_4-C_3	1.555	$C_5C_4C_3$	109 42	$C_{10}-C_9$	1.498	$C_{11}C_{10}C_9$	110 00
C_3-C_2	1.534	$C_4C_3C_2$	113 12	C_9-C_8	1.520	$C_{10}C_9C_8$	114 12
C_2-N_1	1.479	$C_3C_2N_1$	109 24	C_8-N_5	1.502	$C_9C_8N_5$	109 42
C_2-C_1	1.539	$N_1C_2C_1$	109 18	C_8-C_7	1.579	$N_5C_8C_7$	108 6
C_1-O_1	1.262	$C_3C_2C_1$	112 30	C_7-O_3	1.251	$C_9C_8C_7$	110 12
C_1-O_2	1.256	$C_2C_1O_1$	115 12	C_7-O_4	1.253	$C_8C_7O_3$	116 24
		$C_2C_1O_2$	118 54			$O_3C_7O_4$	126 00
		$O_1C_1O_2$	125 36			$C_8C_7O_4$	117 24

one trans). There are also wide variations in the angle between the plane of the carboxyl group and that of C_α, C_β, C_γ, although this is the same for the two independent molecules of a compound. The position is rather different for the angle between the plane of the guanidine group and that of C_γ, C_δ, N_ϵ, since this angle differs even as between the two independent types of molecule in the hydrobromide and hydrochloride monohydrates [87].

In every case the arginine molecule is present in zwitterion form. Only the halides have three hydrogen bonds to the amino group and five to the guanidine group (both groups in protonized form), whereas the dihydrate has only the guanidine group in protonized form, the other group being present as neutral NH_2 [87].

Lysine (α, ε-Diaminocaproic Acid)

$H_2NCH_2CH_2CH_2CH_2CH(NH_2)COOH$

This occurs in nearly all proteins of animal origin, but it is absent (or present in very small amounts) in plant proteins. It was isolated in 1889 from casein, but it was at first taken to be a diamine, and the true chemical formula was not established until 1902 [18-20].

TABLE 48. Atomic Parameters in the Structure of L-Lysine Hydrochloride Dihydrate [Temperature Factors of the Form $T_j = \exp-(B_{11}h^2 + B_{22}k^2 + B_{33}l^2 + B_{12}hk + B_{13}hl + B_{23}kl)$]

Atom	x	y	z	$10^4 \cdot B_{11}$	$10^4 \cdot B_{22}$	$10^4 \cdot B_{33}$	$10^4 \cdot B_{12}$	$10^4 \cdot B_{13}$	$10^4 \cdot B_{23}$
O_1	0.3568	0.0596	-0.3904	174	64	180	5	73	13
O_2	0.1038	0.1213	-0.2893	153	64	255	36	22	60
C_1	0.2503	0.0761	-0.2483	130	47	167	-20	44	7
C_2	0.3026	0.0339	-0.0066	113	38	182	6	35	1
N_1	0.1982	0.0855	0.1560	148	43	193	14	66	-1
C_3	0.2721	-0.0793	-0.0082	183	33	251	6	58	2
C_4	0.3126	-0.1293	0.2250	186	42	235	9	39	20
C_5	0.2920	-0.2430	0.2044	218	42	261	-6	78	20
C_6	0.3034	-0.2927	0.4389	183	47	202	-1	33	10
N_2	0.2800	-0.4027	0.4158	174	43	229	-14	46	34
$O_1(H_2O)_1$	0.1283	0.3156	0.5342	584	68	402	-59	278	22
$O_2(H_2O)_2$	0.2945	0.2897	0.1563	465	52	368	1	153	46
Cl^-	0.2077	0.5003	-0.1137	163	56	242	8	87	30

Atom	x	y	z	B_j, $\overset{\circ}{A}^2$	Atom	x	y	z	B_j, $\overset{\circ}{A}^2$
$H_1(N_1)$	0.197	0.153	0.140	4	$H_{11}(C_6)$	0.425	-0.282	0.543	5
$H_2(N_1)$	0.263	0.076	0.317	4	$H_{12}(C_6)$	0.221	-0.258	0.543	5
$H_3(N_1)$	0.086	0.066	0.111	4	$H_{13}(C_5)$	0.414	-0.277	0.121	5
$H_4(N_2)$	0.172	-0.421	0.291	4.5	$H_{14}(C_5)$	0.169	-0.270	0.110	5
$H_5(N_2)$	0.380	-0.427	0.411	4.5	$H_{15}(C_4)$	0.227	-0.107	0.357	4.5
$H_6(N_2)$	0.263	-0.430	0.565	4.5	$H_{16}(C_4)$	0.438	-0.114	0.307	4.5
$H_7(H_2O)_1$	0.122	0.358	0.650	8	$H_{17}(C_3)$	0.140	-0.091	-0.096	4.0
$H_8(H_2O)_1$	0.129	0.248	0.612	8	$H_{18}(C_3)$	0.367	-0.115	-0.109	4.0
$H_9(H_2O)_2$	0.291	0.354	0.063	7.0	$H_{19}(C_2)$	0.420	0.041	0.058	3.0
$H_{10}(H_2O)_2$	0.247	0.314	0.302	7.0					

In 1956 the compound was first examined by x-ray methods [88], which gave the space group and cell parameters of the hydrochloride dihydrate. In 1959 Raman deduced a rough structure for L-lysine hydrochloride dihydrate and determined the absolute configuration of the L-lysine molecule [89]. A complete study was performed by Wright and Marsh in 1961 [90].

The crystals were grown by slow evaporation of an aqueous solution of L-lysine HCl at room temperature. Space group $P2_1$, cell parameters a = 7.492 Å; b = 13.320 Å; c = 5.879 Å; β = 97°47.4'; ρ_{meas} = 1.250 g/cm³; Z = 2 (lys · HCl · 2H₂O); ρ_X = 1.259 g/cm³.

The trial model was derived by minimizing the three-dimensional sharpened function of the interatomic vectors. The refinement (including the parameters of the hydrogen atoms) was performed by three-dimensional least squares and via a three-dimensional differential electron-density synthesis. This gave R(hkl) as 5.7%, with the distances to ~0.015 Å and the angles to ~40'. Table 48 gives the atomic parameters.

Figure 58 shows the molecule; the distances and angles agree well with the mean values found for other amino acids, and the mean C–H (1.06 Å) and O–H (0.98 Å) (Table 49) agree within the limits of error of measurement with the standard values of 1.10 and 0.97 Å, respectively [91], whereas the mean N–H distance (0.94 Å) is less than the standard 1.01 Å.

The distribution of the hydrogen atoms between the amino and carboxyl groups (Fig. 59) implies that the molecule is present as a zwitterion in which both amino groups have the NH_3^+ form and the carboxyl has the COO^- form. There are two planar groups: 1) carboxyl, the aliphatic chain together with N_2; these two planes meet at 71.4°, N_1 deviating from the plane of the carboxyl group by 0.446 Å.

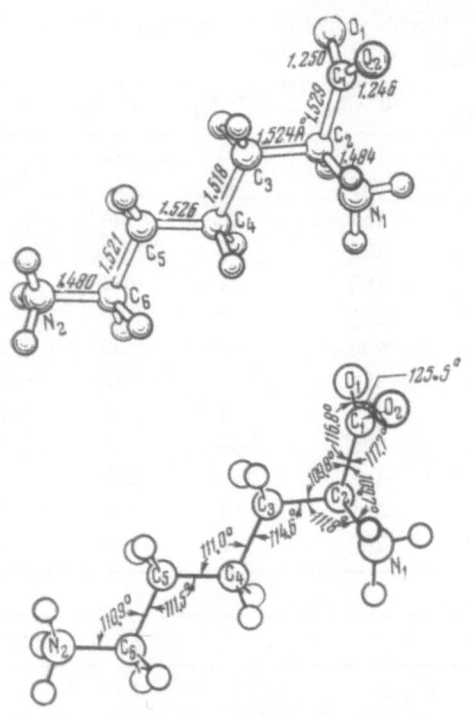

Fig. 58. Form of the molecule in L-lysine hydrochloride dihydrate.

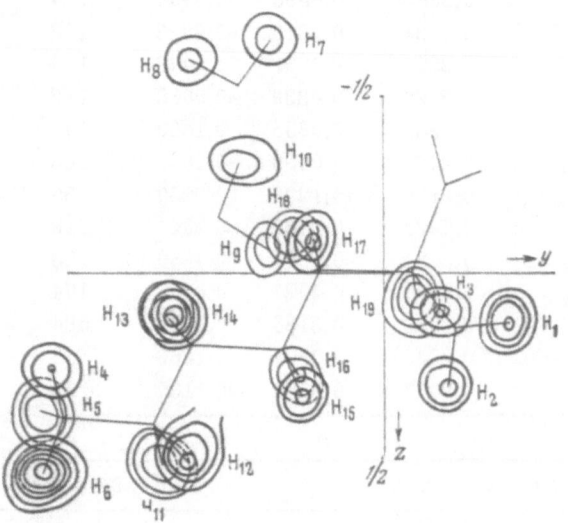

Fig. 59. Projection of sections of the differential synthesis used to locate the hydrogen atoms in the structure of L-lysine hydrochloride dihydrate.

TABLE 49. Lengths of Bonds Containing Hydrogen Atoms in the Structure of L-Lysine Hydrochloride Dihydrate

N_1-H_1 0.91 Å	$O_1(H_2O)-H_7$ 0.89 Å	C_4-H_{14} 1.07 Å
N_1-H_2 1.01	$O_1(H_2O)-H_7$ 1.00	C_4-H_{15} 1.11
N_1-H_3 0.88	$O_2(H_2O)-H_9$ 1.01	C_4-H_{16} 1.02
N_2-H_4 1.04	$O_2(H_2O)-H_{10}$ 1.02	C_3-H_{17} 1.07
N_2-H_5 0.82	C_6-H_{11} 1.04	C_3-H_{18} 1.09
N_2-H_6 0.97	C_6-H_{12} 1.03	C_2-H_{19} 0.91
	C_6-H_{13} 1.18	

There is a network of N–H...O, N–H...Cl⁻, O–H...O, and O–H...Cl⁻ bonds (Fig. 60), which involve all 10 protons on the nitrogen atoms and on the oxygen atoms of the water molecules. One linked to N_2 is bifurcated:

$$N—H \begin{array}{c} Cl^- \\ \\ O^- \end{array}$$

6. SULFUR AMINO ACIDS

Cysteine (α-Amino-β-Mercaptopropionic Acid)

HSCH₂CH(NH₂)COOH

This is a rather unstable compound; in neutral or acid solution it oxidizes fairly rapidly to cystine [18-20].

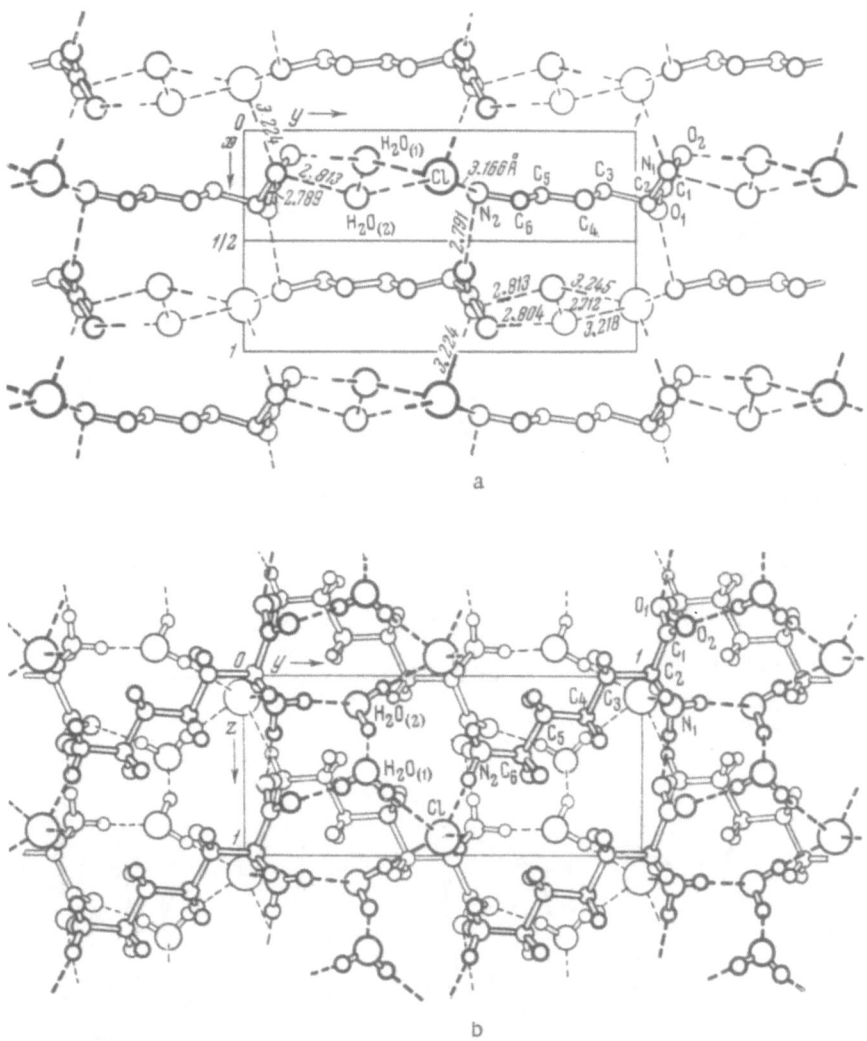

Fig. 60. Schematic representation of the structure of L-lysine hydrochloride dihydrate: (a) view along c axis, (b) view along a axis.

There is no evidence on the crystal structure of cysteine itself, but the configuration of the cysteine residue is known for the structure of S-methyl-L-cysteine sulfoxide [92].

Crystals of this compound were grown from aqueous alcohol; space group $P2_12_12_1$, cell parameters $a = 5.214$ Å; $b = 7.410$ Å; $c = 16.548$ Å; $\rho_{meas} = 1.56$ g/cm³; $Z = 4$; $\rho_X = 1.57$ g/cm³. The model was derived via sign relations for the $F(0kl)$ and an xy Patterson projection. The structure was refined via electron-density projections and also two- and three-dimensional least-squares treatments, the coordinates of the hydrogen atoms being deduced from crystallographic considerations. The minimal $R(hkl)$ was 12.8%, the distances being found to 0.02-0.03 Å and the angles to 1-2°. Table 50 gives the atomic coordinates; Fig. 61 shows the distances and valence angles. Atom C_2 lies in the plane of the carboxyl group, the nitrogen atom deviating from this by 0.185 Å. The C−N bond has a length close to the mean for the C−N bonds of amino acids, whereas C_1−C_2 is lengthened, while C_2−C_3, C_1−O_1, and C_1−O_2 are shortened. It is possible that C_1−O_1 and C_1−O_2 have been somewhat distorted by inexact location of C_1; displacement of C_1 along the C_1−C_2 bond by an amount equal to the probable error increases these bond lengths to 1.26 and 1.24 Å, respectively.

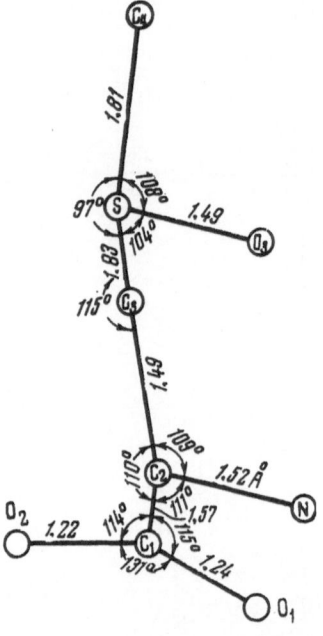

Fig. 61. Form of the molecule in the structure of S-methyl-L-cysteine sulfoxide.

TABLE 50. Atomic Coordinates in the Structure of S-Methyl-L-Cysteine Sulfoxide

Atom	x	y	z
C_1	-0.0388	0.0896	-0.1197
C_2	0.2387	0.0064	-0.1188
C_3	0.2249	-0.1897	-0.1389
C_4	0,4162	-0.5302	-0.1533
O_1	-0.1122	0.1504	-0.0538
O_2	-0.1355	0.0982	-0.1865
O_3	0.6703	-0.2727	-0.0739
N	0.3633	0.0295	-0.0367
S	0,5363	-0.3005	-0.1523
H_1	0.317	-0.073	-0,002
H_2	0.293	0.137	-0.010
H_3	0.550	0.003	-0.041
H_4	0.361	0.071	-0.164
H_5	0.129	-0.253	-0.089
H_6	0.121	-0.201	-0.194

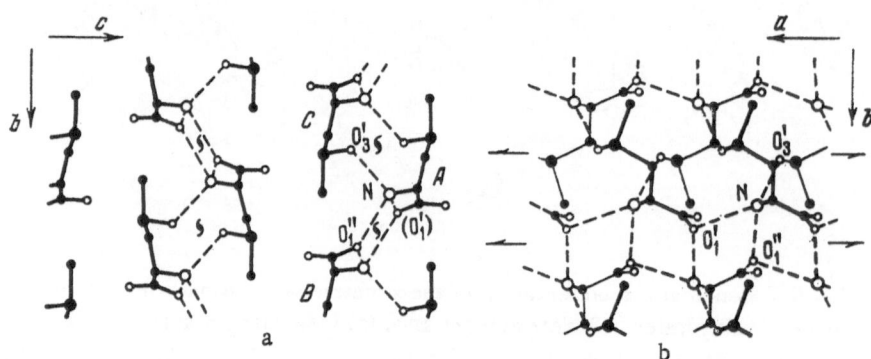

Fig. 62. Schematic representation of the structure of S-methyl-L-cysteine sulfoxide.

Figure 62 shows the packing of the molecules, which are linked together by $N-H...O_1'$ (2.89 Å), $N-H...O_2''$ (2.81 Å), and $N-H...O_3'$ (2.83 Å) bonds into double layers parallel to (100). There are only intermolecular forces between the double layers; this explains the good cleavage.

Cystine (β, β'-Dithiodi-α-Aminopropionic Acid)

$[SCH_2CH(NH_2)COOH]_2$

This occurs widely, especially in the proteins of horn, hair, and feathers. The disulfide (S—S) bridge causes it to play an important part in the secondary and tertiary structures of protein molecules [5, 16].

It was discovered in 1810 in urinary calculi, which are produced by a particular metabolic disorder [18-20].

The study of the crystal structure began in 1956 [93, 94] and was completed in 1959 [95]. The crystals were grown from a warm solution of L-cystine in 10% ammonia by the addition of warm dilute acetic acid until a precipitate appeared, the solution then being cooled. Space group $P6_122$, cell parameters $a = 5.422$ Å; $c = 56.275$ Å; $\rho_{meas} = 1.677$ g/cm^3; $Z = 6$; $\rho_x = 1.671$ g/cm^3.

The model was derived from the three-dimensional Patterson function; the structure was refined from two three-dimensional electron-density distributions and one three-dimensional differential synthesis, the last giving the coordinates of the hydrogen atoms (Table 51). $R(hkl)$ was 12.3%, while the distances were determined to ∼0.015 Å and the angles to ∼1°.

The bond lengths and valence angles are given in Table 52 (see Fig. 63 for notation)

The molecule consists of two cysteine halves transformed one to the other by a twofold axis parallel to [11$\bar{2}$0]. The atoms $C_2N_1C_3O_1O_2$ of each asymmetric part lie in one plane, with the C_1-C_2 bond at 106°10' to this. A zwitterion form is indicated by the disposition of the hydrogen atoms.

The planar $C_2N_1C_3O_1O_2$ groups of all molecules lie nearly at right angles to the c axis and are linked via N−H...O (2.789 Å; 2.809 Å) into layers resembling the single layers in α-glycine (Fig. 64a). The disulfide bridges link the layers into pairs, the pairs being linked via N−H...O (2.865 Å) (Fig. 64b).

The hydrochloride [96-98] and hydrobromide [99-101] have also been studied.

Corsmit et al. [96] made a preliminary study of the hydrochloride, while Steinrauf et al. [97, 98] performed a more complete study. The crystals were grown by slow evaporation of a solution of L-cystine in dilute hydrochloric acid. Space group C2, cell parameters $a = 18.61$ Å; $b = 5.25$ Å; $c = 7.23$ Å; $\beta = 103.6°$; $\rho_{meas} = 1.520$ g/cm^3; $Z = 2$ (cys · 2HCl); $\rho_x = 1.515$ g/cm^3.

The model was found from Patterson projections by vector convergence, refinement being by electron-density projection, including differential projections, the coordinates of the hydrogen atoms being deduced from

TABLE 51. Atomic Parameters in the Structure of
L-Cystine (Temperature Factors of the Form
$T_j = \exp - B_j \sin^2 \theta / \lambda^2$)

Atom	x	y	z	B_j, Å^2
S	0.19866	0.03138	0.41278	T_S
C_1	0.0886	0.7310	0.39270	3.28
C_2	0.0515	0.7836	0.36697	3.20
C_3	0.7967	0.8301	0.36231	3.20
N_1	0.3165	0.0337	0.35686	3.06
O_2	0.8493	0.0732	0.35613	3.37
O_2	0.5586	0.6200	0.36578	3.81

$T_S = \exp - (0.04139h^2 + 0.04555k^2 + 0.000166l^2 + 0.04555hk)$

Atom	x	y	z
H_1	0.223	0.675	0.3923
H_2	0.926	0.562	0.3961
H_3	0.010	0.620	0.3563
H_4	0.493	0.015	0.3538
H_5	0.383	0.215	0.3667
H_6	0.283	0.985	0.3390

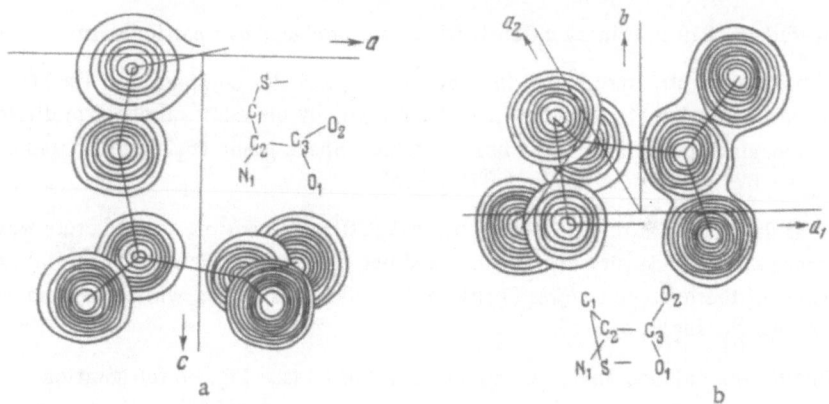

Fig. 63. Configuration of the asymmetric part of the cystine molecule:
(1) view along twofold axis parallel to [11$\bar{2}$0], (b) view along [0001].
The lines are at intervals of 1 e/$\overset{\circ}{A}^3$ for C, N, and O, but at 5 e/$\overset{\circ}{A}^3$ for S.

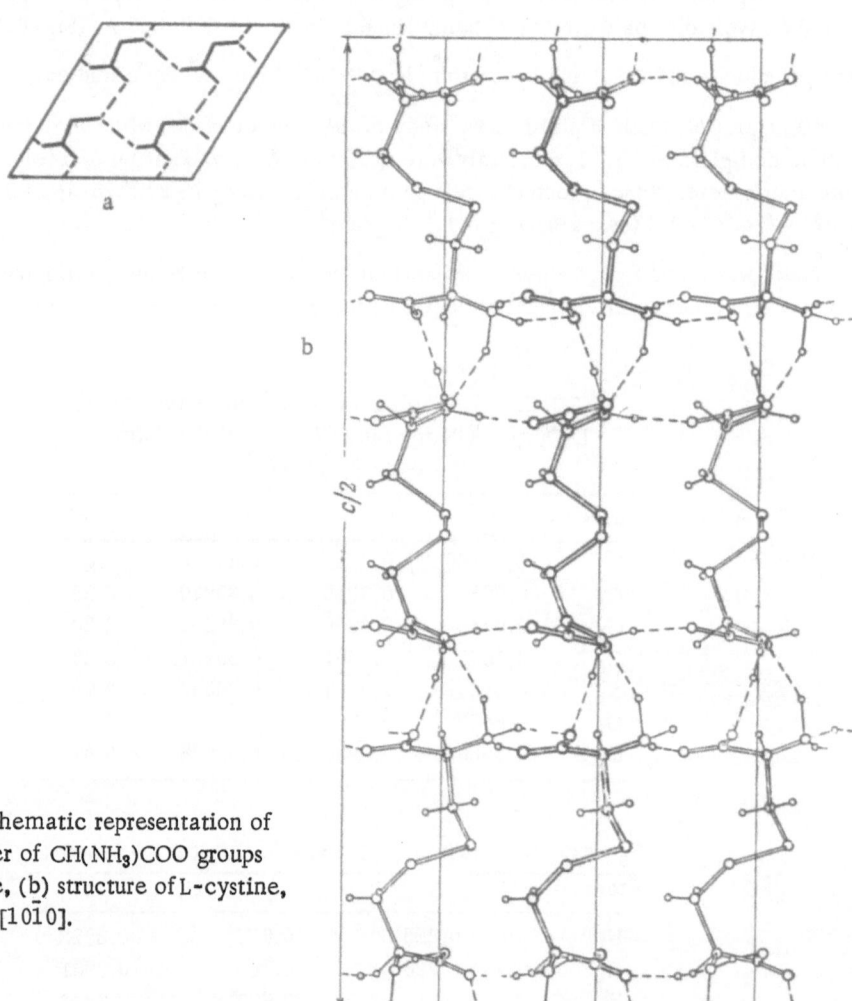

Fig. 64. Schematic representation of
(a) one layer of CH(NH₃)COO groups
in L-cystine, (b) structure of L-cystine,
view along [10$\bar{1}$0].

TABLE 52.　Interatomic Distances and Valence Angles
in the Structure of L-Cystine

$S-S'$	2.032 Å	$S'-S-C_1$	114°30'
$S-C_1$	1.820	$S-C_1-C_2$	116 12
C_1-C_2	1.509	$C_1-C_2-C_3$	114 13
C_2-C_3	1.543	$C_1-C_2-N_1$	111 54
C_2-N_1	1.511	$N_1-C_2-C_3$	108 32
C_3-O_1	1.250	$C_2-C_3-O_1$	117 54
C_3-O_2	1.238	$C_2-C_3-O_2$	115 20
N_1-H_4	1.04	$O_1-C_3-O_2$	126 48
N_1-H_5	1.02	$C_2-N_1-H_4$	119
N_1-H_6	1.07	$C_2-N_1-H_5$	112
		$C_2-N_1-H_6$	100

Angle $C_1'-S'-S-C_1$ 106°

TABLE 53.　Atomic Coordinates in the Structure of
L-Cystine Hydrochloride

Atom	x	y	z
C_1	0.0642	0.7660	0.2406
C_2	0.1465	0.6790	0.3048
C_3	0.1552	0.4670	0.4418
N	0.1765	0.6100	0.1388
O_1	0.1924	0.2750	0.4292
O_2	0.1195	0.4790	0.5772
S	0.0003	0.5000	0.1415
Cl^-	0.1532	0.0890	0.8848
H_1	0.067	0.910	0.100
H_2	0.057	0.900	0.367
H_3	0.188	0.850	0.377
H_4	0.150	0.460	0.033
H_5	0.172	0.825	0.033
H_6	0.233	0.460	0.167
H_7	0.133	0.360	0.667

crystallographic considerations and also from the latter projections (Table 53). The final R(hk0) and R(h0l) were 7.9%.

The principal interatomic distances and valence angles are given in Table 54.

The asymmetric parts of the molecule have a structure analogous to that found in cystine itself, but the general configuration is different, being as found in the peptide N,N'-diglycyl-L-cystine [102] (Fig. 65).

The distribution of the hydrogen atoms corresponds to the form $[SCH_2CH(NH)_3]^+$ for the molecule. The main links between molecules are via $N-H \ldots Cl^-$ and $O-H \ldots Cl^-$ bonds (Table 54).

The structure of L-cystine hydrobromide has been described [99-101]; the data given here are from [100, 101]. These crystals were not isomorphous with those of the hydrochloride. Space group $P2_12_12_1$, cell parameters a = 17.85 Å; b = 5.35 Å; c = 7.48 Å; ρ_{meas} = 1.870 g/cm^3; Z = 2 (cys · 2HBr); ρ_X = 1.869 g/cm^3.

TABLE 54. Interatomic Distances and Valence Angles
in the Structure of L--Cystine Hydrochloride

S—S'	2.044 Å	∠S'—S—C_1	103.8°
S—C_1	1.865	S—C_1—C_2	112.8
C_1—C_2	1.561	C_1—C_2—C_3	111.5
C_2—C_3	1.474	C_1—C_2—N	111.2
C_2—N	1.482	C_2—C_3—O_1	122.7
C_3—O_1	1.238	C_2—C_3—O_2	118.1
C_3—O_2	1.307	N—C_2—C_3	110.9
O_2—H...Cl^-	2.98	O_1—C_3—O_2	119.1
N'—H...Cl^-	3.08	Angle C_1—S'—S—C_1 = 79°2'	
N"—H...Cl^-	3.27		
N'''—H...Cl^-	3.25		

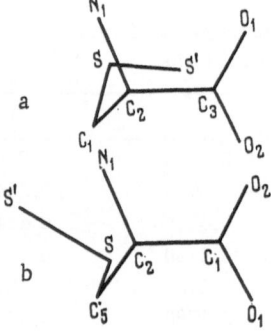

Fig. 65. Projection of the asymmetric part of the cystine molecule on the plane of the C_α(N)COO group in: (a) L-cystine, (b) N, N'-diglycyl-L-cystine.

The entire analysis was based on the h0l and hk0 intensities. The model was deduced via Zachariasen's statistical relations; refinement was by differential electron-density projections with anisotropic temperature factors, which gave R(h0l) = R(hk0) = 7.1%. The interatomic C—X(C, N, O) distances were determined to ~0.3 Å and the S—S distances to ~0.01 Å. Table 55 gives the coordinates.

The structure is analogous to that of the hydrochloride, in spite of the absence of isomorphism; the molecule has almost the same configuration in both. The bond lengths and valence angles agree within the error of measurement, while the packing is governed by an analogous system of hydrogen bonds (Table 56).

Methionine (α-Amino-γ-Methylthiobutyric Acid)

$H_3CSCH_2CH_2CH(NH_2)COOH$

This amino acid is one of the principal sources of sulfur in proteins. It was observed in casein by Muller in 1922, and its chemical structure was established in 1928 [18-20].

Mathieson [103, 104] examined the structure in 1951. The crystals were grown from a solution of the racemate in aqueous alcohol by slow evaporation. It was found that two distinct modifications were produced. The name α form was given to crystals having space group $P2_1/a$ and cell parameters as follows: a = 9.76 Å; b = 4.70 Å; c = 16.70 Å; β = 102°; ρ_{meas} = 1.34 g/cm³; Z = 4; ρ_x = 1.33 g/cm³; while the name β form was given to ones having space group I2/a and parameters a = 9.94 Å; b = 4.70 Å; c = 33.40 Å; β = 106.6°; ρ_{meas} = 1.34 g/cm³; Z = 8; ρ_x = 1.33 g/cm³.

TABLE 55. Atomic Coordinates in the Structure of L-Cystine Hydrobromide

Atom	x	y	z	Atom	x	y	z
C_1	0.0606	0.8820	0.2050	$H_1(C_1)$	0.059	0.010	0.080
C_2	0.1394	0.7905	0.2320	$H_2(C_1)$	0.042	0.040	0.307
C_3	0.1457	0.5745	0.3608	$H_3(C_2)$	0.165	0.950	0.277
N	0.1795	0.7210	0.0641	$H_4(N)$	0.162	0.560	0.973
O_1	0.1795	0.3845	0.3233	$H_5(N)$	0.165	0.900	0.980
O_2	0.1128	0.6097	0.5091	$H_6(N)$	0.232	0.680	0.145
S	0.9963	0.6254	0.1350	$H_7(O_2)$	0.125	0.450	0.625
Br^-	0.1579	0.2046	0.7984				

Parameters of the Temperature Factors $T_j = \exp-[(A + C \cos^2\psi) \sin^2 \theta / \lambda^2)]$

	A_{hk0}	C_{hk0}	ψ_{hk0}	A_{h0l}	C_{h0l}	ψ_{h0l}
C_1	1.70	0.00	—	1.70	0.00	—
C_2	1.70	0.00	—	1.70	0.00	—
C_3	1.70	0.00	—	1.70	0.00	—
N	1.70	0.00	—	1.70	0.00	—
O_1	3.50	0.00	—	3.50	0.00	—
O_2	3.50	0.00	—	3.50	0.00	—
Br^-	2.04	0.80	-40.0	1.65	1.10	-54.0
S	1.19	1.41	0.0	1.10	1.40	14.4

TABLE 56. Interatomic Distances and Valence Angles
in the Structure of L-Cystine Hydrobromide

S—S'	2.024 Å	S'—S—C_1	103.9°
S—C_1	1.862	S—C_1—C_2	111.9
C_1—C_2	1.506	C_1—C_2—C_3	113.9
C_2—C_3	1.509	C_1—C_2—N	114.7
C_2—N	1.493	C_2—C_3—O_1	122.0
C_3—O_1	1.215	C_2—C_3—O_2	114.2
C_3—O_2	1.269	N—C_2—C_3	108.1
O_2—H...Br^-	3.17	O_1—C_3—O_2	123.8
N'—H...Br^-	3.28	Angle C_1'—S'—S—C_1 =	
N''—H...Br^-	3.42	90°	
N'''—H...Br^-	3.41		

TABLE 57. Coordinates of the Main Atoms in the Structure of α-DL-Methionine

Atom	x	y	z
S	0.328	-0.056	0.128
O_1	0.007	-0.094	0.378
O_2	0.174	-0.342	0.435
N	0.364	0.053	0.406
C_1	0.125	-0.144	0.392
C_2	0.218	0.028	0.353
C_3	0.225	-0.096	0.268
C_4	0.311	0.093	0.224
C_5	0.164	0.047	0.068

TABLE 58. Coordinates of the Main Atoms in the Structure of β-DL-Methionine

Atom	x	y	z
S	0.017	0.172	0.059
O_1	-0.270	0.192	0.187
O_2	-0.091	-0.078	0.216
N	0.099	0.306	0.203
C_1	-0.142	0.122	0.194
C_2	-0.049	0.306	0.175
C_3	-0.051	0.169	0.131
C_4	0.038	0.336	0.109
C_5	0.151	0.350	0.043

TABLE 59. Interatomic Distances and Valence Angles in the Structure of:

α-DL-methionine				β-DL-methionine			
C_5-S	1.77 Å	$\angle C_5-S-C_4$	100°	C_5-S	1.78 Å	$\angle C_5-S-C_4$	100°
$S-C_4$	1.79	$S-C_4-C_3$	111	$S-C_4$	1.80	$S-C_4-C_3$	109
C_4-C_3	1.51	$C_4-C_3-C_2$	111	C_4-C_3	1.54	$C_4-C_3-C_2$	113
C_3-C_2	1.55	$C_3-C_2-C_1$	111	C_3-C_2	1.58	$C_3-C_2-C_1$	108
C_2-C_1	1.47	$C_2-C_1-O_1$	120	C_2-C_1	1.52	$C_3-C_1-O_1$	118
C_2-N	1.52	$C_2-C_1-O_2$	119	C_2-N	1.50	$C_2-C_1-O_2$	120
C_1-O_1	1.28	$O_1-C_1-O_2$	121	C_1-O_1	1.27	$O_1-C_1-O_2$	122
C_1-O_2	1.21	C_3-C_2-N	119	C_1-O_2	1.21	C_3-C_2-N	109
$N-H..._1O_1$	2.92	C_1-C_2-N	112	$N-H..._1O_1$	2.82	C_1-C_2-N	110
$N-H..._2O_1$	2.59	$O_1-N-_2O_1$	117	$N-H..._2O_1$	2.80	$_1O_1-N-_2O_1$	113
$N-H..._1O_2$	2.80	$C_2-N-_1O_1$	111	$N-H..._1O_2$	2.78	$C_2-N-_1O_1$	108
		$C_2-N-_2O_1$	105			$C_2-N-_2O_1$	110
		$C_1-O_2-_1N$	126			$C_2-N-_1O_2$	108
		$C_1-O_1-_2N$	125			$C_1-O_2-_1N$	128
		$C_2-N-_1O_2$	106			$C_1-O_1-_2N$	136
		$C_1-O_1-_3N$	109			$C_1-O_1-_3N$	103

The model for α-methionine was deduced from Harker sections (y = 0 and y = 1/2), with refinement via xz and yz electron-density projections. The final R(h0l) and R(0kl) was 21%, the distances being found to ~0.04 Å. Table 57 gives the coordinates of the main atoms.

The model for the β form was deduced on the basis of similarity to the α form, because two of the cell parameters are identical. The refinement was again via electron-density projections (Table 58), which gave R as 21% for h0l and 23% for 0kl.

The bond lengths and valence angles of both forms are given in Table 59.

All of these are close to the values found for other amino acids. The C—C bond lengths alternate along the chain. In the α and β forms:

1) The carboxyl groups are asymmetric, because the oxygen atoms differ in hydrogen bonding;

2) The group $NC_2C_1O_1O_2$ is not planar, with the N atom 0.70 Å out of the plane of the carboxyl group, which exceeds the deviations usually found;

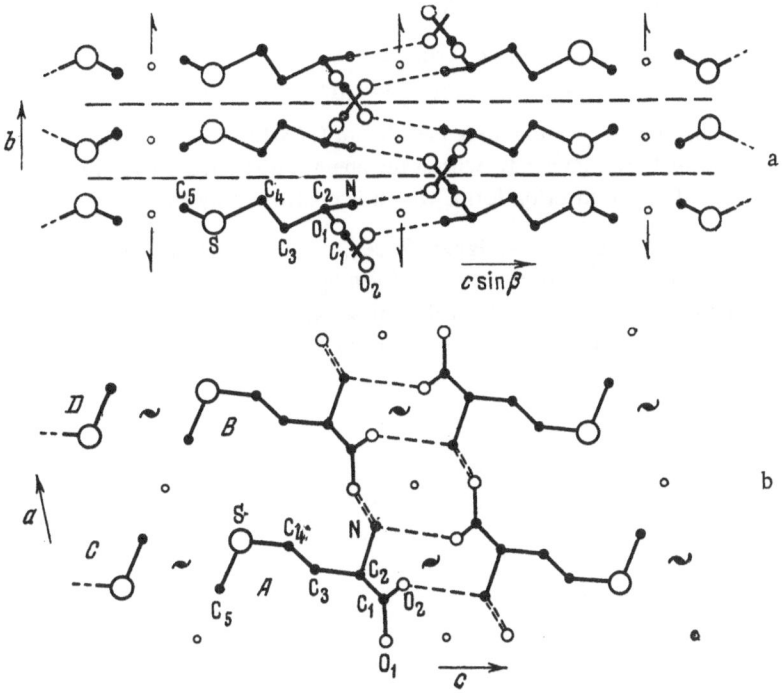

Fig. 66. Schematic representation of the structure of α-DL-methionine: (a) view along a axis, (b) view along b axis.

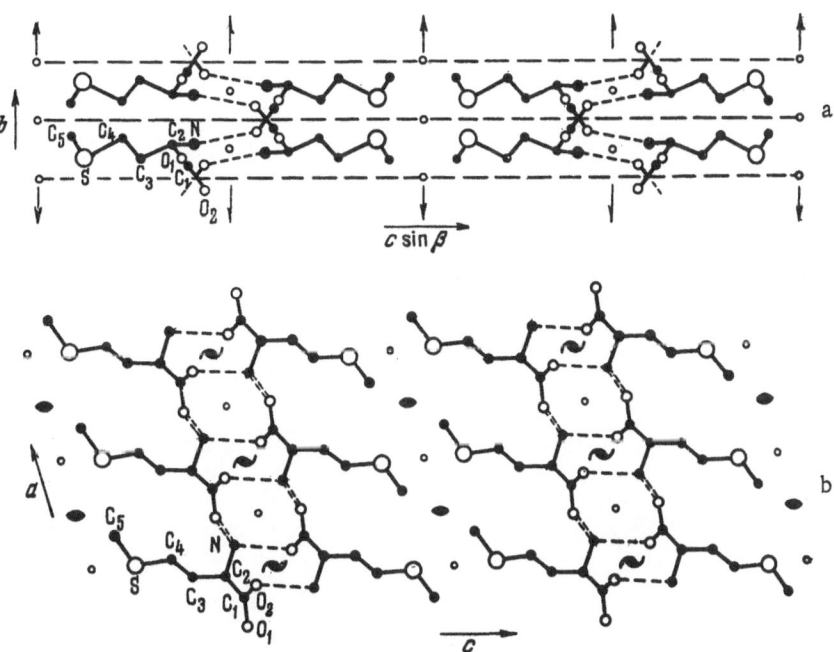

Fig. 67. Schematic representation of the structure of β-DL-methionine: (a) view along a axis, (b) view along b axis.

3) C_1, C_2, C_3, C_4, and S form a nearly planar zigzag chain.

The main difference between the α and β forms lies in the orientation of the C_5-S bond relative to the plane of the hydrocarbon chain: C_5 lies in that plane in β-methionine but projects from it in α-methionine.

Both forms have double layers of molecules parallel to (100) (Figs. 66 and 67), each double layer having a complete internal system of N−H...O bonds, with a tetrahedral disposition with respect to each other and the C−N bond, so the molecule has a zwitterion form, which is confirmed by the absence of O−H...O bonds.

The double layers are linked by van der Waals forces between the terminal groups. The difference in the orientation of the C_5-S bonds give rise to two ways of forming a balance system of intermolecular forces; the double layers in α-methionine are related by twofold screw axes and centers of symmetry, while those in β-methionine are related by simple twofold axes and centers.

CHAPTER II

CRYSTAL STRUCTURES OF AROMATIC
AND HETEROCYCLIC AMINO ACIDS

7. AROMATIC AMINO ACIDS

Phenylalanine (α-Amino-β-Phenylpropionic Acid)

 — CH₂CH(NH₂)COOH

This is one of the essential amino acids and occurs in nearly all proteins; it was first isolated from natural products in 1879, and the chemical formula was established in 1882 [18-20].

In 1931 Bernal determined the space group and cell parameters of the D form [23]. In 1961 Marsh and Glusker established the structure of the phenylalanine residue in the tripeptide glycylphenylalanylglycine [105]. A complete structure study of phenylalanine was performed in 1963 [106-108].

The crystals of L-phenylalanine were grown from aqueous hydrochloric acid by slow evaporation; space group $P2_12_12_1$, cell parameters a = 27.68 Å; b = 6.98 Å; c = 5.34 Å; ρ_{meas} = 1.247 g/cm³; Z = 4 (phal·HCl).

The model was derived from the modified three-dimensional Patterson function by fourfold minimization; refinement was via electron-density projection, three-dimensional least squares, and two three-dimensional electron-density distributions, the coordinates of the hydrogen atoms being deduced from crystallographic considerations, and also in part from a differential xy projection. This gave R(hk$l$) as 13.8%, the distances being found to ~0.02 Å. Table 60 gives the parameters.

Figure 68 shows the interatomic distances and valence angles. The mean C—C bond length (1.54 Å) in the aliphatic part of the molecule coincides with the generally accepted length for a single C—C bond. The mean C—C bond length in the benzene ring is 1.37 Å, which is close to the mean for glycylphenylalanylglycine (1.38 Å) [105] and L-tyrosine·HCl (1.387 Å) [112].

The carboxyl group is clearly unsymmetrical, the hydrogen atom being joined to O_1; the C_1-O_1 bond (1.34 Å) is close to the length for a single C—O bond, while C_1-O_2 (1.17 Å) is close to the length for C=O.

The amino group has the form NH_3^+, so the molecule is present as the positive ion $C_6H_5CH_2CH(NH_3)^+$. COOH. The C_2-N distance (1.48 Å) is close to the standard value for a single C—N bond.

There are two planar groups. One consists of the benzene ring plus C_3, while the other consists of the carboxyl group and C_2. The nitrogen atom lies very nearly in the latter plane (0.058 Å away), the C_2-N bond lying at 2° to that plane.

The configuration of the molecule differs from that of the residue in gly-phal-gly [105]; one may be converted to the other by rotation of the ring around the C_2-C_3 bond.

The packing is determined by a complete system of N—H...Cl⁻, N—H...O, and O—H...Cl⁻ hydrogen bonds, together with van der Waals interactions. The molecules and chloride ions lie in double layers parallel.

81

TABLE 60. Atomic Parameters in the Structure of
L-Phenylalanine Hydrochloride * (Temperature Factors
of the form $T_j = \exp - B_j \sin^2 \theta / \lambda^2$)

Atom	x	y	z	B_j
Cl^-	0.0561	0.1511	0.9768	3.76
C_1	0.0645	0.7012	0.3188	3.23
C_2	0.0686	0.5819	0.5509	3.56
C_3	0.1206	0.5449	0.6465	3.20
C_4	0.1534	0.4398	0.4506	3.43
C_5	0.1765	0.5404	0.2678	5.81
C_6	0.2075	0.4543	0.0984	6.57
C_7	0.2134	0.2588	0.1134	7.05
C_8	0.1888	0.1544	0.3054	7.76
C_9	0.1587	0.2470	0.4724	5.42
O_1	0.0860	0.8711	0.3588	4.93
O_2	0.0457	0.6591	0.1316	3.69
N	0.0424	0.4027	0.4956	3.71
$H_1(O_1)$	0.065	0.917	0.777	
$H_2(N)$	0.056	0.347	0.623	
$H_3(N)$	0.041	0.332	0.359	
$H_4(N)$	0.011	0.440	0.543	
$H_5(C_2)$	0.048	0.641	0.297	
$H_6(C_3)$	0.140	0.660	0.322	
$H_7(C_3)$	0.114	0.585	0.144	
$H_8(C_5)$	0.170	0.695	0.745	
$H_9(C_6)$	0.228	0.535	0.043	
$H_{10}(C_7)$	0.236	0.185	0.014	
$H_{11}(C_8)$	0.194	0.005	0.682	
$H_{12}(C_9)$	0.138	0.165	0.386	

* The B_j for the H atoms have been taken as $B_j + 1$, in
which B_j corresponds to the heavy atom to which the
H is joined.

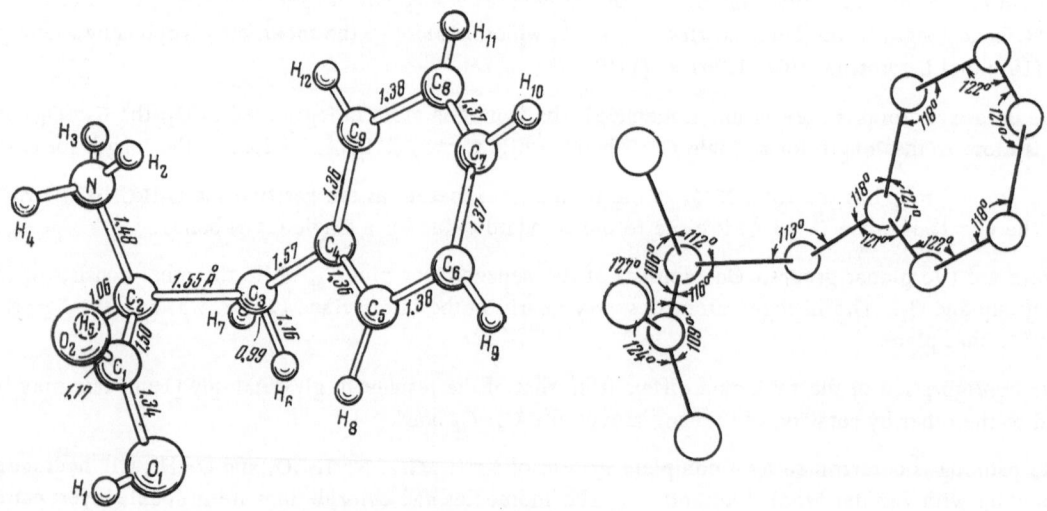

Fig. 68. Structure of L-phenylalanine molecule in the crystal of the hydrochloride.

to (100) (Fig. 69a), the disposition of the double layers in the crystal being determined only by van der Waals forces (Fig. 69b), which is responsible for the good cleavage on (100).

Tyrosine (α-Amino-β-Hydroxyphenylpropionic Acid)

$$HO - \langle\!\!\!\bigcirc\!\!\!\rangle - CH_2CH(NH_2)COOH$$

This is one of the commoner amino acids and was discovered in casein in 1846. Its chemical formula was established in 1883 [18-20]. The structure has been examined [109-112] on the isomorphous hydrochloride and hydrobromide, both of which have space group $P2_1$, the cell parameters being similar:

L-tyrosine $\cdot$ HBr   L-tyrosine $\cdot$ HCl
$a = 11.38$ Å $a = 11.07$ Å
$b = 9.12$ Å $b = 9.03$ Å
$c = 5.175$ Å $c = 5.09$ Å
$\beta = 91.2°$ $\beta = 91.8°$
$\rho_{meas} = 1.64$ g/cm^3 $\rho_{meas} = 1.42$ g/cm^3
$Z = 2$ (tyr $\cdot$ HBr) $Z = 2$ (tyr $\cdot$ HCl)
$\rho_X = 1.62$ g/cm^3 $\rho_X = 1.42$ g/cm^3

The model for the hydrobromide was deduced by superposition of the xy projection of the differential Patterson function and also by xz projection of the ordinary Patterson function [111]. Refinement was by ordinary and differential electron-density projections and also two-dimensional least squares with anisotropic temperature factors. The final R was 10.5% for hk0, 10.9% for h0l, and 10.7% for 0kl. The distances were determined to 0.04-0.05 Å and the angles to 3—4°. Table 61 gives the coordinates of the main atoms.

Figure 70 shows the molecular structure. All distances except C_1—O_1 agree with the results for other amino acids, as do the angles. The C_1—O_1 distance (1.42 Å) is larger than the usual 1.37 Å for a carbon atom in an aromatic ring and a hydroxyl oxygen.

The atoms in the benzene ring are coplanar within the accuracy of the measurements, while O_1 and C_7 deviate from the plane by 0.087 Å and 0.042 Å, respectively. A further planar group in the molecule consists of the carboxyl group and C_8. The nitrogen atom of the amino group deviates from this plane by 0.62 Å.

TABLE 61. Coordinates of the Main Atoms in the
Structure of L-Tyrosine Hydrobromide

Atom	x	y	z
C_1	0.093	0.025	0.361
C_2	0.080	0.129	0.151
C_3	0.972	0.130	0.020
C_4	0.880	0.028	0.084
C_5	0.902	0.932	0.302
C_6	0.005	0.928	0.432
C_7	0.760	0.033	0.938
C_8	0.660	0.086	0.090
C_9	0.675	0.247	0.170
O_1	0.195	0.028	0.518
O_2	0.643	0.343	0.010
O_3	0.738	0.274	0.372
N	0.550	0.067	0.940
Br$^-$	0.4123	0.2500	0.4368

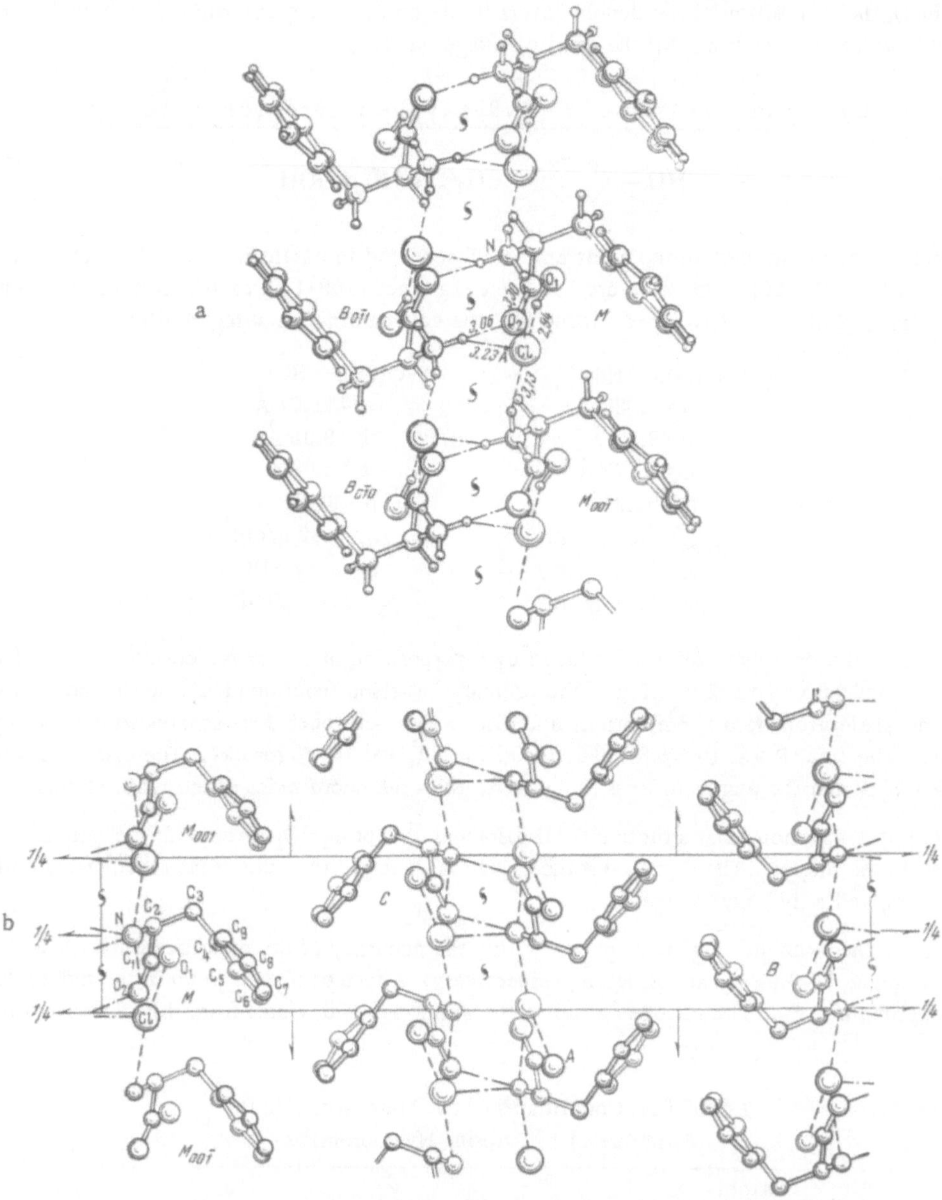

Fig. 69. Structure of L-phenylalanine hydrochloride: (a) formation of hydrogen bonds in the double layers, (b) packing of double layers in crystals. View along b axis.

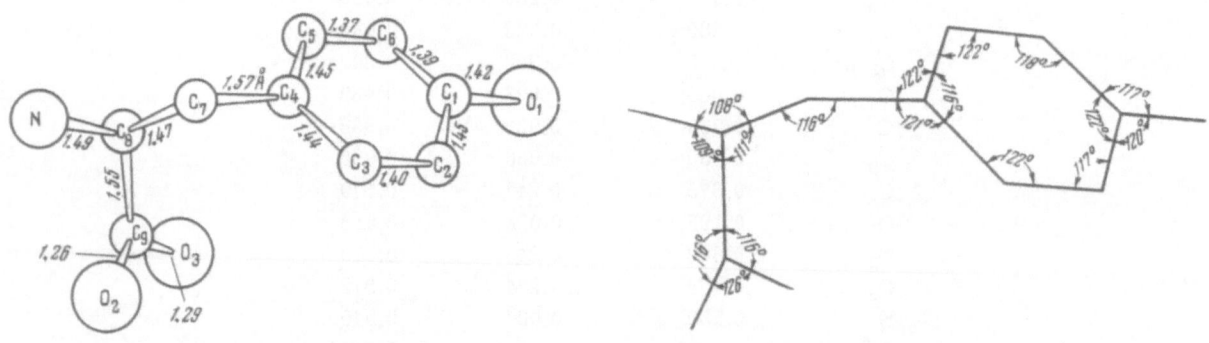

Fig. 70. Form of the molecule in the structure of L-tyrosine hydrobromide.

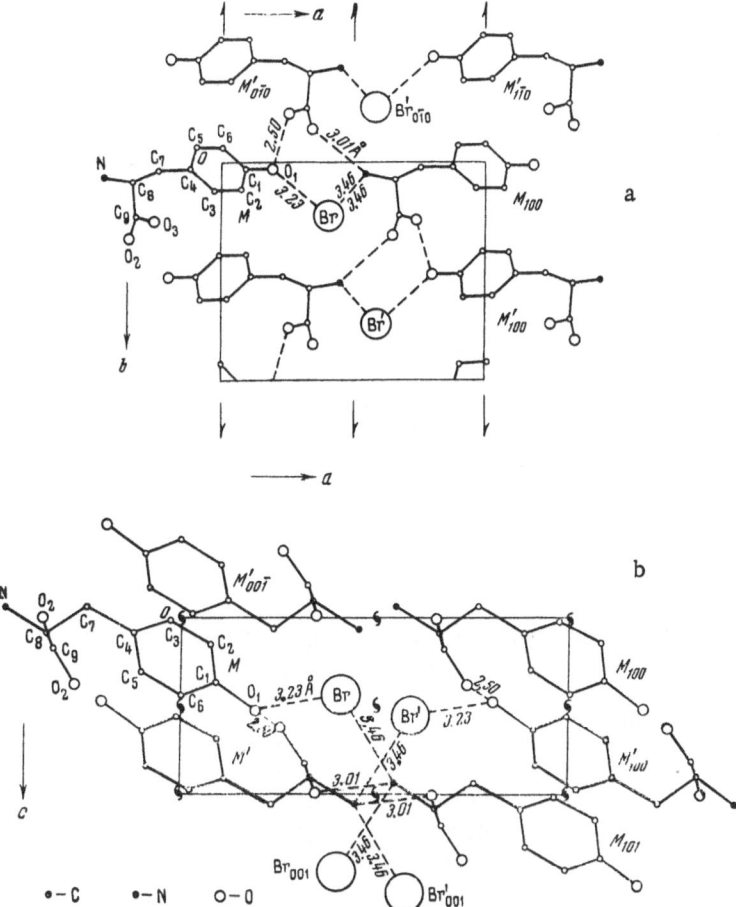

Fig. 71. Schematic representation of the structure of L-tyrosine hydrobromide: (a) view along c axis, (b) view along b axis.

Figure 71 illustrates the packing in the crystal. The molecules extend along the longest axis (the a axis) and are localized in layers almost parallel to (010). Molecules denoted by M and M' belong to separate layers; the coupling within and between layers is via N$-$H...O, N$-$H...Br$^-$, O$-$H...O, and O$-$H...Br$^-$ bonds. The N$-$H...O and N$-$H...Br bonds are formed from hydrogen in the NH$_3^+$ group, while O$-$H...Br$^-$ and O$-$H...O are formed by hydrogen atoms in the hydroxyl and carboxyl groups.

The model for L-tyrosine hydrochloride was derived by taking the coordinates of O, N, and C as equal to those for the refined structure of the hydrobromide, while the coordinates of the chloride ion were deduced from Patterson projections. This model was refined by least squares with anisotropic temperature factors and also by differential electron-density projection, the coordinates of the hydrogen atoms being deduced from crystallographic considerations, and also via differential syntheses. This gave R as 13% for h0l and 10.3% for hk0. The distances were found to 0.02-0.03 Å and the angles to 2-3°[112]. Table 62 gives the coordinates.

Figure 72 shows the molecule in the structure of L-tyrosine hydrochloride. The bond lengths are closer to the standard values than those found for the hydrobromide. The molecular configuration is much the same in the two compounds; O$_1$ deviates from the plane of the ring by 0.080 Å, and C$_7$ by 0.075 Å, while the nitrogen atom is 0.71 Å from the plane of the carboxyl group. The distribution of the hydrogen atoms corresponds to the form $HOC_6H_4CH_2CH(NH_3)^+COOH$.

Figure 73 shows that the mode of packing is analogous to that for the hydrobromide.

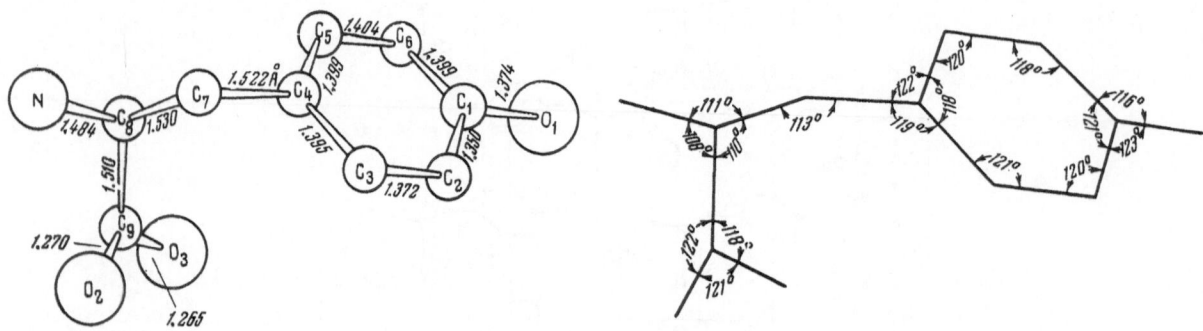

Fig. 72. Molecular shape in the structure of L-tyrosine hydrochloride.

TABLE 62. Atomic Coordinates in the Structure of
L-Tyrosine Hydrochloride

Atom	x	y	z
C_1	0.0990	0.035	0.362
C_2	0.0815	0.127	0.157
C_3	0.9730	0.128	0.019
C_4	0.8785	0.033	0.088
C_5	0.8970	0.941	0.304
C_6	0.0080	0.940	0.445
C_7	0.7620	0.035	0.922
C_8	0.6530	0.087	0.075
C_9	0.6690	0.247	0.155
O_1	0.2020	0.034	0.520
O_2	0.6390	0.353	0.003
O_3	0.7230	0.273	0.373
N	0.5400	0.074	0.912
Cl^-	0.4130	0.250	0.405
$H_1(O_1)$	0.26	0.10	0.51
$H_2(C_2)$	0.13	0.21	0.12
$H_3(C_3)$	0.96	0.21	0.90
$H_4(C_7)$	0.77	0.13	0.88
$H_5(C_5)$	0.86	0.84	0.36
$H_6(C_6)$	0.01	0.84	0.56
$H_7(C_7)$	0.74	0.97	0.80
$H_8(C_8)$	0.63	0.01	0.13
$H_9(O_3)$	0.75	0.36	0.46
$H_{10}(N)$	0.50	0.15	0.05
$H_{11}(N)$	0.50	0.15	0.73
$H_{12}(N)$	0.56	0.99	0.82

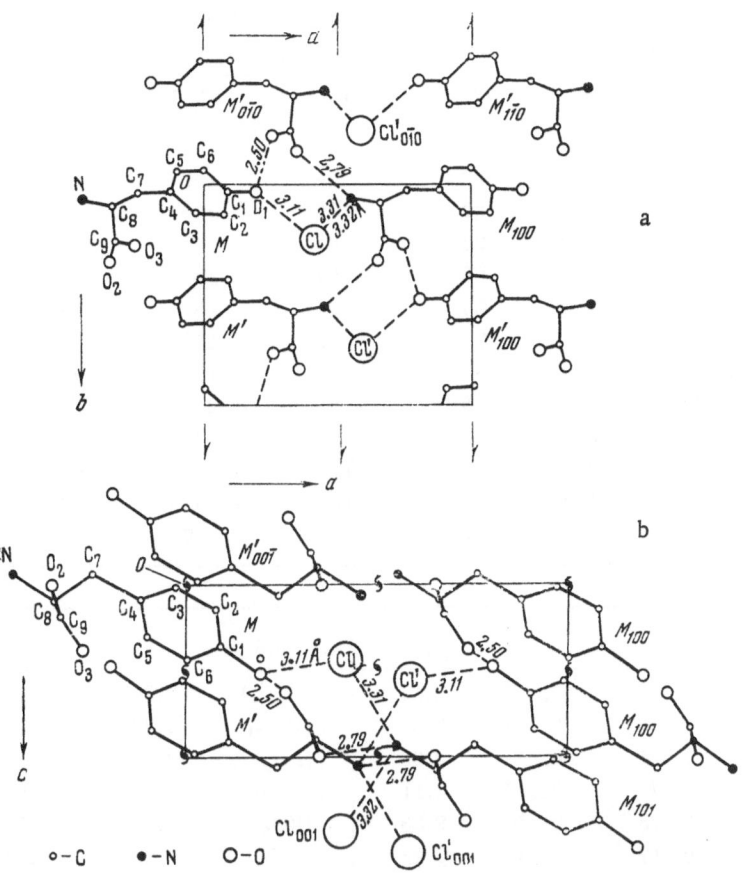

Fig. 73. Schematic representation of the structure of L-tyrosine hydrochloride: (a) view along c axis, (b) view along b axis.

8. HETEROCYCLIC AMINO ACIDS

Tryptophan [α-Amino-β-(3-Indolyl)propionic Acid]

$$\text{—CH}_2\text{CH(NH}_2)\text{COOH}$$
NH

This occurs in many proteins, but usually in small amounts; it was discovered in 1901 in casein, and its chemical structure was established in 1907 [18-20].

The crystal structure has not yet been examined, the only evidence being for the residue in the dipeptide glycyltryptophan·$2H_2O$ [113] (Fig. 74) and from the unit-cell parameters of DL-tryptophan dihydrochloride [61]: $a = 5.30$ Å; $b = 8.61$ Å; $c = 16.20$ Å; $\alpha = 103°$; $\beta = 107°$; $\gamma = 98°$; $Z = 2$. Space group P1 or P$\bar{1}$.

Histidine (α-Amino-β-Imidazopropionic Acid)

$$\text{—CH}_2\text{CH(NH}_2)\text{COOH}$$
HN   N

This is a common amino acid, which occurs in especially large amounts in hemoglobin. It was isolated from protein in 1896 by Kossel, and its chemical formula was finally established in 1911 [18-20]. The struc-

TABLE 63. Atomic Parameters in the Structure Histidine Hydrochloride Monohydrate [Temperature Factors of the Form $T_j = \exp-(B_{11}h^2 + B_{22}k^2 + B_{33}l^2 + B_{12}hk + B_{13}hl + B_{23}kl)$]

Atom	x	y	z	$10^4 \cdot B_{11}$	$10^4 \cdot B_{22}$	$10^4 \cdot B_{33}$	$10^4 \cdot B_{12}$	$10^4 \cdot B_{13}$	$10^4 \cdot B_{23}$
C_1	0.3645	0.1939	0.5044	36	53	114	5	16	23
C_2	0,4020	0.1169	0.3240	20	63	150	10	-9	-13
C_3	0.4513	0.2237	0.1900	20	76	155	9	22	-20
C_4	0.4016	0.3625	0.1321	23	45	125	-21	13	-17
C_5	0.3414	0.5251	0.9337	29	97	187	-12	-19	28
C_6	0.3680	0.4766	0.2392	26	74	187	-7	14	23
N_1	0.3305	0.0384	0.2173	22	60	148	2	9	-34
N_2	0.3836	0.3964	0.9397	33	95	100	-21	13	-10
N_3	0.3301	0.5734	0.1123	35	54	213	-2	4	20
O_1	0.2857	0.1836	0.5413	26	96	171	6	32	-34
O_2	0.4204	0.2628	0.6055	35	147	136	-26	0	-77
O_3	0.0802	0,1001	0.4083	36	187	208	29	3	112
Cl^-	0.1758	0.2276	0.0299	34	69	126	16	-5	4

Atom	x	y	z	B_j, $\overset{\circ}{A}^2$	Atom	x	y	z	B_j, $\overset{\circ}{A}^2$
$H(C_2)$	0.446	0.039	0.369	2.17	$H''(N_1)$	0.350	0.000	0.087	2.23
$H(C_3)$	0.472	0.163	0.075	2.36	$H'''(N_1)$	0.305	-0.504	0.296	2.23
$H(C_3)$	0.506	0,251	0.268	2.36	$H(N_2)$	0.402	0.333	0.821	2.57
$H(C_5)$	0,321	0.575	0.809	3.11	$H(N_3)$	0.299	0.670	0.150	2.80
$H(C_6)$	0.370	0.485	0.384	2.53	$H(O_3)$	0.111	0.142	0.297	4.17
$H'(N_1)$	0.280	0.112	0.187	2.23	$H'(O_3)$	0.223	0.147	0.406	4.17

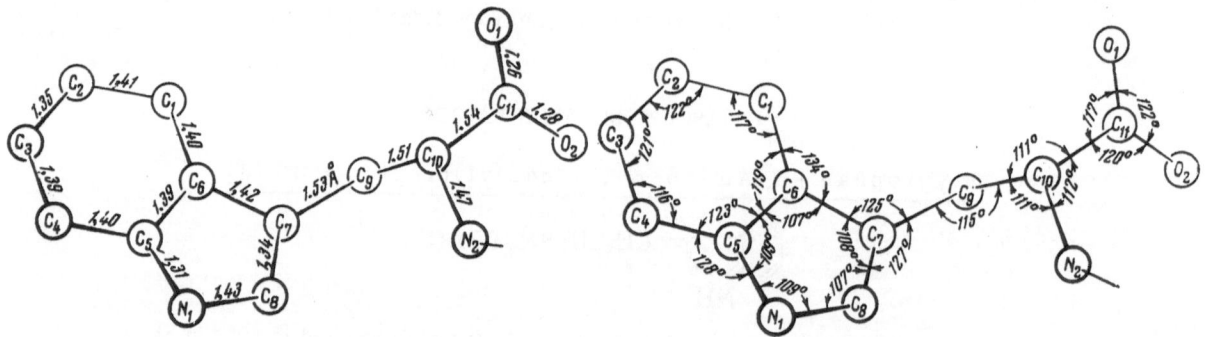

Fig. 74. Structure of the tryptophan residue in glycyl-L-tryptophan · $2H_2O$.

ture was established on crystals of the hydrochloride monohydrate [114-116] and the hydrated zinc salts [117, 118].

Crystals of the hydrochloride monohydrate were grown by slow evaporation from aqueous solution; space group $P2_12_12_1$, cell parameters a = 15.36 Å; b = 8.92 Å; c = 6.88 Å; ρ_{meas} = 1.485 g/cm³; Z = 4 (his · HCl · H_2O); ρ_x = 1.477 g/cm³.

The model was derived by vector convergence from the modified three-dimensional Patterson function; refinement by least squares, the coordinates of the hydrogen atoms being deduced from crystallographic considerations (Table 63). The final R(hkl) was 7.6%; the distances were determined to ~0.010 Å.

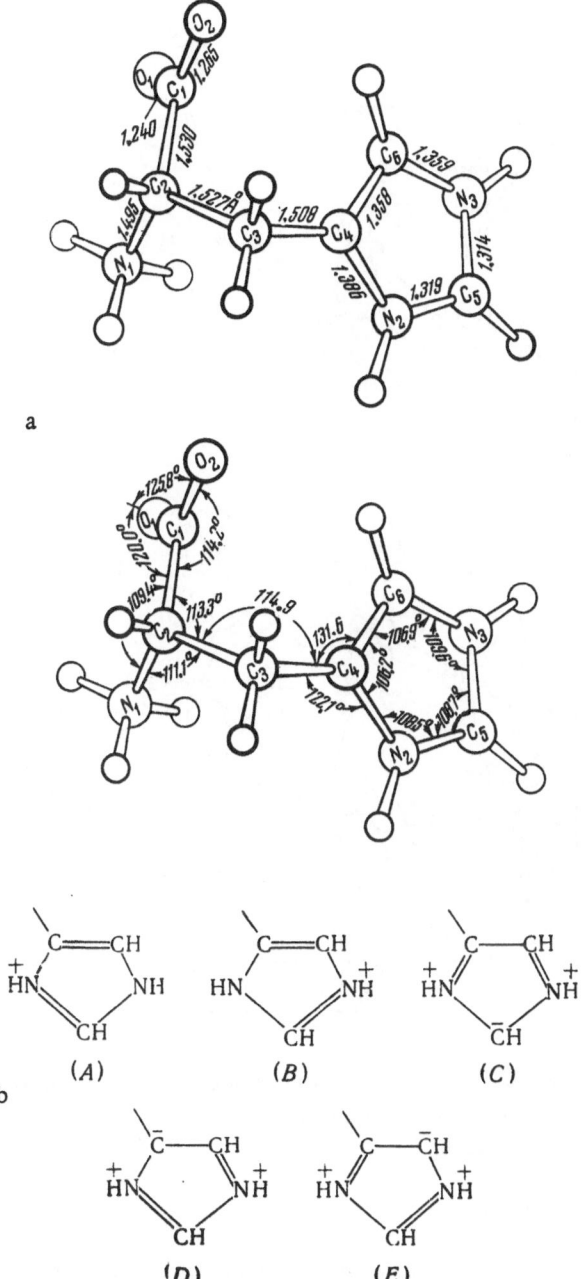

Fig. 75. (a) Shape of the molecule in the structure of histidine hydrochloride monohydrate, (b) possible resonance forms of imidazole ring.

Figure 75a shows the valence angles and intramolecular interatomic distances, all of which agree well with the analogous quantities summarized in Tables 72-75.

The distribution of the hydrogen atoms allows one to represent the imidazole ring as several resonance forms (Fig. 75b), whose relative weights have been calculated [115] from the relation of length to character for the C—C and C—N bonds. All calculations were performed after the first stage of refinement, when R(hkl)

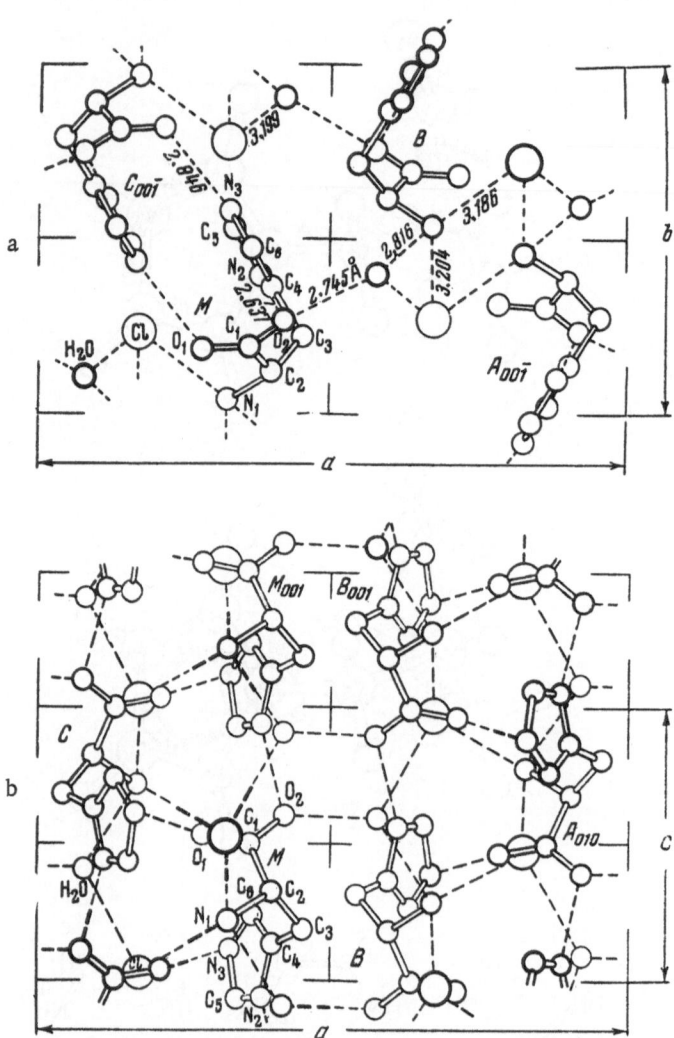

Fig. 76. Schematic representation of the structure of histidine hydrochloride monohydrate: (a) view along c axis, (b) view along b axis.

was 12%, so the C—C and C—N bond lengths were somewhat less accurate than those shown in Fig. 75a. The results agreed well with the expected values: 40% A, 32% B, 23.6% C, 5% D, and 0% E.

There are two planar groups: C_2 with the carboxyl group and C_3 with the imidazole ring. The carboxyl group is almost symmetrical.

The molecules are linked by N—H...O, N—H...Cl⁻, O—H...Cl⁻, and O—H...O bonds (Fig. 76), with three hydrogen bonds per amino-group nitrogen, these going to Cl⁻ and oxygen atoms of water molecules. The latter also have two bonds to the oxygen of the carboxyl groups, where they act as hydrogen donors. The imidazole nitrogen atoms form one hydrogen bond each to the carboxyl oxygen.

Zinc histidinate pentahydrate crystals [118] were produced by slow cooling of a saturated solution of DL-histidine and zinc sulfate; space group C2/c, cell parameters $a = 16.41$ Å; $b = 14.755$ Å; $c = 10.99$ Å; $\beta = 129.6°$; $\rho_{meas} = 1.504$ g/cm³; $Z = 4$ (2 his·Zn·5H₂O); $\rho_x = 1.502$ g/cm³.

The position of the zinc atoms was determined from projections of the sharpened Patterson function, while the coordinates of the light atoms in the anions were determined by superposition of the sharpened inter-atomic-vector function. The structure was refined by differential electron-density synthesis and by least squares, the coordinates of the hydrogen atoms being deduced from crystallographic considerations, and from the differential synthesis. The final R(hk$l$) was 10.5%.

The Zn$-$X(N, O) distances were determined to $\sim$0.012 Å and C$-$N(O) to $\sim$0.020 Å; the valence angles were determined to $\sim$1.3°. Table 64 gives the atomic parameters.

TABLE 64. Atomic Parameters in the Structure of Zn(DL-histidine)$_2 \cdot$5H$_2$O [Temperature Factors of the Form $T_j = \exp-(B_{11}h^2 + B_{22}k^2 + B_{33}l^2 + B_{23}kl + B_{31}hl + B_{12}hk)$]

Atom	x	y	z	$10^4 \cdot B_{11}$	$10^4 \cdot B_{22}$	$10^4 \cdot B_{33}$	$10^4 \cdot B_{23}$	$10^4 \cdot B_{31}$	$10^4 \cdot B_{12}$
Zn	0.0000	0.0372	0.2500	28.8	20.8	92.5	0.0	59.8	0.0
C_1	0.1733	0.0767	0.6328	47.6	22.0	105.6	15.5	109.9	-11.7
C_2	0.1908	-0.0097	0.5775	44.2	21.7	82.6	-23.6	87.3	0.1
C_3	0.2626	0.0094	0.5354	31.3	37.3	89.5	-17.1	75.1	-5.3
C_4	0.2286	0.0891	0.4302	37.6	31.8	85.4	-26.3	93.2	-9.2
C_5	0.1217	0.1842	0.2346	63.4	25.0	107.4	-9.6	136.0	-8.9
C_6	0.2887	0.1553	0.4341	50.5	41.1	113.3	-44.6	113.7	-32.9
N_1	0.0894	-0.0481	0.4403	36.4	24.3	101.9	21.3	85.5	-15.9
N_2	0.1217	0.1079	0.2989	36.9	25.9	76.1	-8.9	83.7	-1.0
N_3	0.2197	0.2135	0.3116	87.1	33.6	155.9	-33.2	186.9	-36.8
O_1	0.0821	0.1071	0.5572	50.4	29.9	176.5	-71.6	125.2	-15.6
O_2	0.2534	0.1141	0.7554	71.3	28.9	91.3	-40.3	101.1	-20.5
O_3	0.1557	0.2152	0.8665	122.7	46.5	284.7	13.4	299.6	16.6
O_4	0.0344	0.4097	0.0148	59.7	107.6	213.8	-68.5	83.7	49.1
O_5	0.0000	0.3424	0.7500	95.8	51.9	154.6	0.0	151.3	0.0

Atom	x	y	z	B_j, Å^2	Atom	x	y	z	B_j, Å^2
$H_1(C_2)$	0.234	-0.052	0.680	3.5	$H_9(O_3)$	0.192	0.200	0.825	6.0
$H_2(N_1)$	0.120	-0.108	0.425	3.5	$H_{10}(O_4)$	0.108	0.400	0.088	6.0
$H_3(N_1)$	0.028	-0.072	0.433	3.5	$\frac{1}{2}H_{11}(O_3)$	0.220	0.247	0.962	6.0
$H_4(C_3)$	0.338	0.000	0.579	3.5	$\frac{1}{2}H_{12}(O_3)$	0.100	0.267	0.825	6.0
$H_5(C_3)$	0.215	-0.045	0.400	3.5	$\frac{1}{2}H_{13}(O_5)$	0.050	0.300	0.762	6.0
$H_6(C_5)$	0.052	0.217	0.135	3.5	$\frac{1}{2}H_{14}(O_5)$	0.008	0.465	0.005	6.0
$H_7(N_3)$	0.230	0.273	0.295	3.5	$\frac{1}{2}H_{15}(O_4)$	0.025	0.384	-0.080	6.0
$H_8(C_6)$	0.374	0.161	0.518	3.5	$\frac{1}{2}H_{16}(O_4)$	0.020	0.380	0.839	6.0

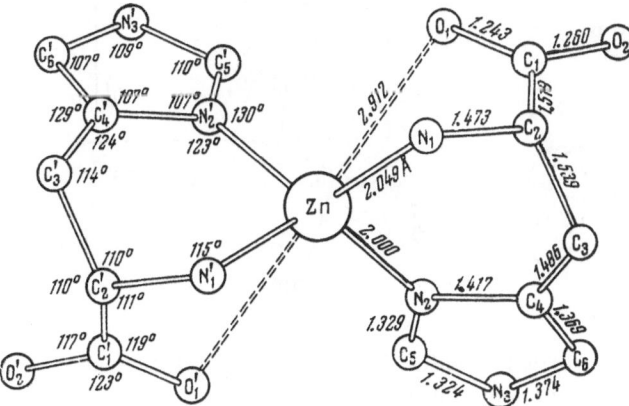

Fig. 77. Form of a single complex in the structure of Zn(DL-histidine)$_2 \cdot$5H$_2$O, seen along twofold axis.

The structure consists of Zn(his)$_2$ complexes (Fig. 77), which are linked together by hydrogen bonds via the water molecules. The zinc ion has a tetrahedral environment of covalently linked nitrogen atoms in the ring and in amino groups. There are also two weak bonds from the zinc to the carboxyl groups.

In this structure the histidine residues differ from those in histidine hydrochloride by rotation of the imidazole ring around the C_3–C_4 bond, which is due to the need to form a tetrahedral array around the zinc. The configuration is otherwise the same in both compounds, namely two planar groups: the carboxyl group with C_2 (nitrogen atom of amino group projecting 0.027 Å) and the imidazole ring with C_3.

Figure 78 shows the system of hydrogen bonds linking the complexes. The lengths of these bonds are given in Table 65.

Crystals of zinc histidinate dihydrate [117] were grown by slow evaporation of an aqueous solution; space group $P4_12_12$, cell parameters a = b = 7.53 Å; c = 30.41 Å; ρ_{meas} = 1.57 g/cm^3; Z = 4 (2·his·Zn·2H$_2$O); ρ_x = 1.57 g/cm^3.

The model was derived by the heavy-atom method and refined by least squares, the coordinates of the hydrogen atoms being deduced from crystallographic considerations; R(hkl) was 11.3%, the distances being determined to ~0.02 Å and the angles to ~2°. Table 66 gives the coordinates of the main atoms.

Figure 79 shows the structure of a single complex; as in the previous case, there are two histidine residues joined to the Zn^{2+} via the nitrogen atoms. The organic parts are related by a twofold axis through the zinc atom. The tetrahedral environment is very much distorted by the fairly close (2.79 Å) Zn–O contacts, which arise by electrostatic interaction of Zn^{2+} with COO$^-$. C_3 lies in the plane of the imidazole ring, while C_2 and the carboxyl group are also coplanar, the nitrogen atom projecting 0.256 Å from this plane. The configuration of the residues is otherwise almost as in the other zinc complex, but the mutual disposition in the complex is altered by the different orientation with respect to the twofold axis through the zinc.

Figure 80 shows the lengths of the hydrogen bonds, and the corresponding angles, in the structure.

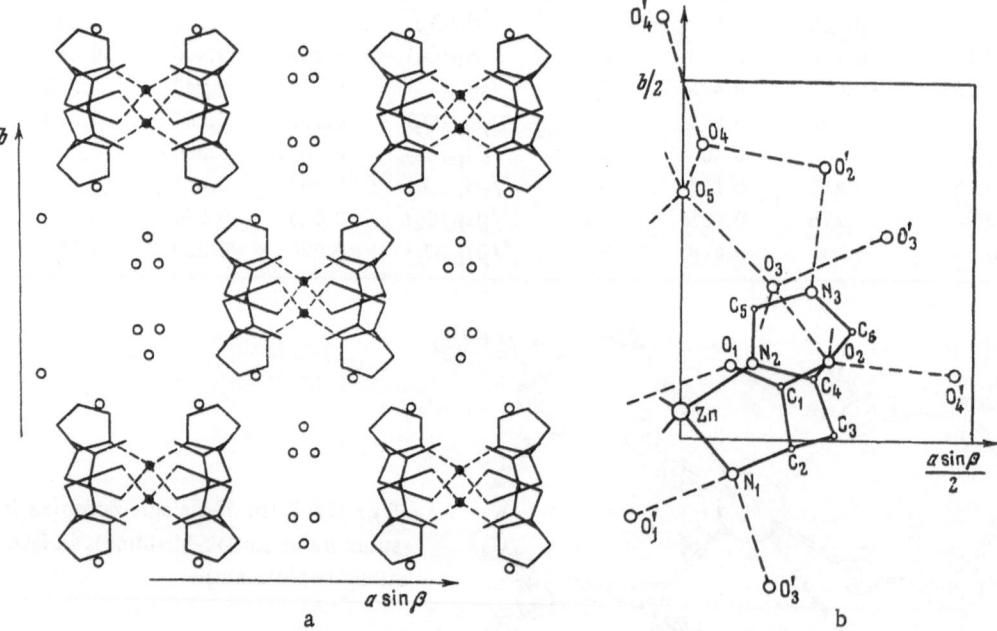

Fig. 78. Schematic representation of the structure of Zn(DL-histidine)$_2$·5H$_2$O: (a) view along c axis, (b) mode of formation of hydrogen bonds.

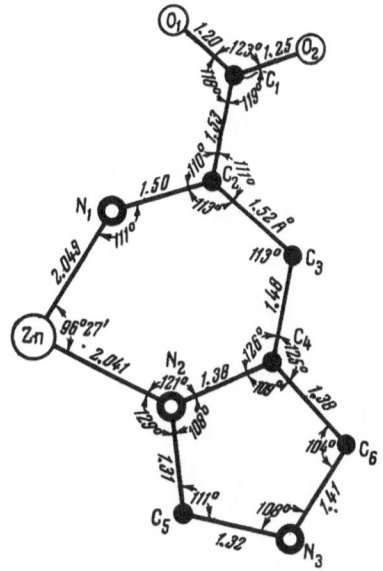

Fig. 79. Form of a single complex in the structure of Zn(L-histidine)$_2 \cdot$ 2H$_2$O (asymmetric part shown).

TABLE 65. Lengths of Hydrogen Bonds in the Structure of Zn(DL-histidine)$_2 \cdot$ 5H$_2$O

N$_3$–H$_7$...O$_2$ ($^1/_2$–x, $^1/_2$–y, 1–z) 2.762 Å
N$_1$–H$_3$...O$_1^{\cdot}$(–x, –y, 1–z) 2.963
N$_1$–H$_2$...O$_3$(x, –y, –$^1/_2$+z) 3.008
O$_3$–H$_9$...O$_2$ 2.964
O$_4$($^1/_2$–x, $^1/_2$–y, 1–z)–H$_{10}$...O$_2$ 2.724
O$_3$...$^1/_2$H$_{11}$...$^1/_2$H$_{11}$...O$_3$($^1/_2$–x, $^1/_2$–y, 2–z) 2.768
O$_3$...$^1/_2$H$_{12}$...$^1/_2$H$_{13}$...O$_5$ 2.742
O$_4$...$^1/_2$H$_{15}$...$^1/_2$H$_{15}$...O$_4$(–x, 1–y, –z) 2.831
O$_4$...$^1/_2$H$_{16}$...$^1/_2$H$_{14}$...O$_5$(x, y, –1+z) 2.771

TABLE 66. Coordinates of the Main Atoms in the Structure of Zn(L-histidine)$_2 \cdot$ 2H$_2$O

Atom	x	y	z	Atom	x	y	z
C$_1$	0.3431	0.0517	0.29981	N$_2$	0.3427	0.4672	0.20545
C$_2$	0.4696	0.0354	0.26104	N$_3$	0.3899	0.3330	0.14476
C$_3$	0.0371	0.3494	0.22250	O$_1$	0.4549	0.2810	0.05165
C$_4$	0.1948	0.3626	0.19671	O$_2$	0.3789	0.1623	0.32883
C$_5$	0.4540	0.4385	0.17487	O(H$_2$O)	0.0648	0.3548	0.34354
C$_6$	0.2238	0.2760	0.15850	Zn	0.1694	0.1694	0.00000
N$_1$	0.1116	0.4186	0.02335				

Proline (α-Pyrrolidine-2-Carboxylic Acid)

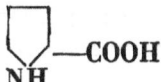

This occurs in many proteins but is present in especially great amounts in the collagen group. The specific structure (one hydrogen atom on the amino group is replaced by the δ carbon of the radical) is respon-

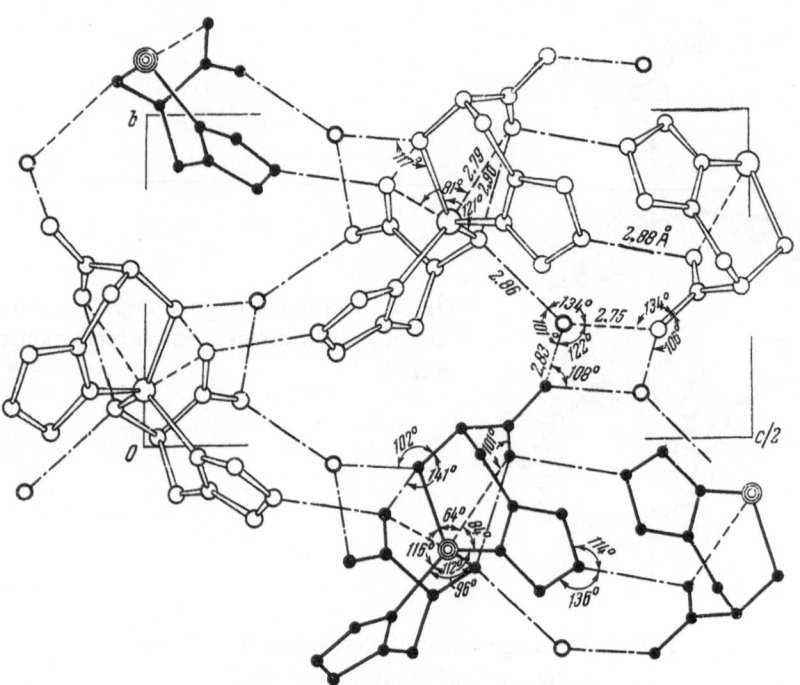

Fig. 80. Packing of complexes in the structure of $Zn(L\text{-histidine})_2 \cdot 2H_2O$.

TABLE 67. Coordinates of the Main Atoms in the Structure of Copper DL-Prolinate Dihydrate

Atom	x	y	z
Cu	0.000	0.000	0.000
O_1	-0.211	0.004	0.185
O_2	-0.311	0.072	0.400
O_3	-0.350	0.085	-0.190
N	0.176	0.085	0.167
C_1	-0.169	0.057	0.303
C_2	0.060	0.104	0.328
C_3	0.011	0.188	0.322
C_4	0.196	0.210	0.219
C_5	0.176	0.156	0.049

sible for the part played in the secondary structure of protein molecules. Willstatter discovered proline by synthesis in 1900; in 1901 Fischer detected it in protein hydrolyzates [18-20].

Wright and Cole [119] in 1949 determined the space group and cell parameters of the L form; in 1952 Mathieson and Welsh [120] derived the structure of copper prolinate dihydrate with low accuracy, while in 1959 Sasisekharan [121] derived the space group and cell parameters for proline monohydrate. The complete structure was determined in 1964 in the protein structure laboratory of this institute [122-124].

Crystals of copper DL-prolinate dihydrate were grown from a solution of copper carbonate in the acid by slow evaporation; space group $P2_1/n$, cell parameters a = 5.62 Å; b = 17.85 Å; c = 7.13 Å; β = 108°; ρ_{meas} = 1.61 g/cm³; Z = 2 (2·pro·Cu·2H$_2$O); ρ_X = 1.59 g/cm³.

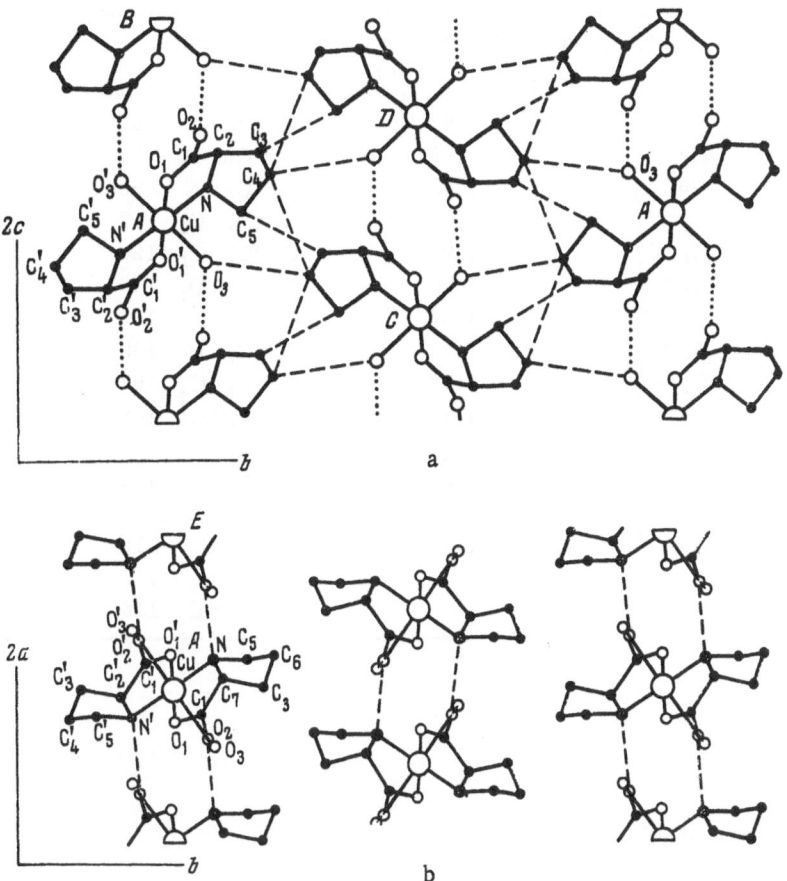

Fig. 81. Schematic representation of the structure of copper DL-prolinate dihydrate: (a) view along a axis, (b) view along c axis.

The preliminary model was derived by trial and error, the structure being refined via electron-density projections, which gave R: R(0kl) = 16.2; R(h0l) = 19.9%; R(hk0) = 15.8%. The distances were determined to ~0.04 Å and the angles to ~4°. Table 67 gives the coordinates of the main atoms.

Figure 81 shows the structure, while the main interatomic distances and valence angles are given below. The structure consists of separate complexes in which the copper atom is surrounded by a slightly distorted octahedron containing two nitrogen atoms and two oxygen atoms from the carboxyl groups of two proline residues together with two oxygen atoms from water molecules. The C−N distances (1.99 Å) and the C−O (carboxyl) distances (2.03 Å) correspond to covalent bonds, while the C−O (water) distances are 2.53 Å. The proline residue contains two planar groups: C_2 with the carboxyl group (nitrogen atom deviating by 0.21 Å) and the ring atoms C_2, C_3, N with C_5 (C_4 projects 0.60 Å from this plane). The carboxyl group and C_4 are in the trans position relative to the plane of the ring.

The complexes are linked via N−H...O and O−H...O bonds into layers parallel to (010) (Fig. 81), there being only intermolecular forces between layers. This explains the good cleavage on (010).

Crystals of L-proline were grown from absolute ethanol by slow evaporation in an atmosphere free from water vapor. Space group $P2_12_12_1$, cell parameters a = 11.55 Å; b = 9.02 Å; c = 5.20 Å; Z = 4.

The preliminary model was derived by minimizing R(hk$l$) by nonlocal search for the minimum in a function of many variables [123, 124]. The structure was refined from the three-dimensional electron density and

TABLE 68. Principal Interatomic Distances and Valence
Angles in the Structure of Copper DL-Prolinate
Dihydrate

$Cu-O_1$	2.03 Å	$\angle N-Cu-O_1$	82°
$Cu-N$	1.99	O_3-Cu-O_1	79
$Cu-O_3$	2.52	O_3-Cu-N	93
C_1-O_1	1.24	$O_1-C_1-O_2$	122
C_1-O_2	1.24	$C_2-C_1-O_1$	120
C_1-C_2	1.50	$C_2-C_1-O_2$	118
C_2-C_3	1.52	$C_1\ C_2-C_3$	112
C_3-C_4	1.50	C_1-C_2-N	108
C_4-C_5	1.52	$C_2-C_3-C_4$	97
C_5-N	1.53	$C_3-C_4-C_5$	109
$N-C_2$	1.52	C_4-C_5-N	96
$N(A)---O_2(E)$	2.86	C_5-N-C_2	108
$O'_3(A)---O'_2(B)$	3.00	$N-C_2-C_3$	108
$O'_3(A)---O_1(E)$	2.94	$Cu-N-C_2$	112
		$Cu-N-C_5$	113
		$Cu-O_1-C_1$	116

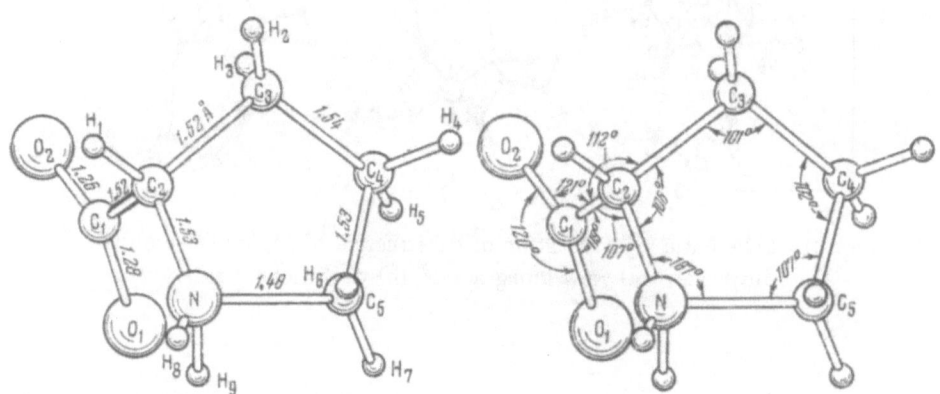

Fig. 82. Form of the molecule in the structure of L-proline.

by least squares, the coordinates of the hydrogen atoms being deduced from crystallographic considerations as well as from Fourier syntheses. The best R(hk$l$) was 16.9%, the distances being to ~0.02 Å. Table 69 gives the atomic parameters.

Figure 82 shows the structure of a single molecule. The C−C, C−O, and C−N bond lengths are close to those found for other amino acids.

The carboxyl group is almost symmetrical, which may mean that the molecule has a zwitterion form. There are two planar groups: C_1, O_1, O_2 (carboxyl) with C_2 (nitrogen atom 0.23 Å from plane) and C_2, C_3, C_5, N (pyrrolidine ring), with C_4 0.60 Å out of this plane. However, the configuration differs from that of the residue in the copper salt, for here C_4 and the carboxyl group lie on the same side of the ring plane (cis form), whereas in the copper salt they are in the trans position.

Figure 83 shows the packing in the crystal. The molecules are linked via $N_M-H_8\ldots O_{1M\,001}$ (2.69 Å) and $N_M-H_9\ldots O_{2B\bar{0}1\bar{0}}$ (2.71 Å), hydrogen bonds into layers parallel to (100), the disposition of the layers being governed by van der Waals forces.

TABLE 69. Atomic Parameters in the Structure of L-Proline (Temperature Factors of the Form $T_j = \exp - B_j \sin^2 \theta / \lambda^2$)

Atom	x	y	z	B_j, Å^2
C_1	0.029	0.268	0.131	3.33
C_2	0.055	0.295	0.415	3.08
C_3	0.153	0.406	0.450	3.82
C_4	0.259	0.305	0.413	4.74
C_5	0.226	0.163	0.557	3.56
O_1	0.099	0.148	0.523	4.10
O_2	0.057	0.144	0.031	3.90
N	-0.023	0.364	-0.001	4.49
$H_1(C_2)$	-0.019(0.04)*	0.331(0.33)	0.531(0.59)	3.08
$H_2(C_3)$	0.152(0.15)	0.448(0.42)	0.648(0.65)	3.82
$H_3(C_3)$	0.151(0.16)	0.487(0.40)	0.291(0.30)	3.82
$H_4(C_4)$	0.333(0.37)	0.364(0.36)	0.497(0.32)	4.74
$H_5(C_4)$	0.268(0.25)	0.260(0.32)	0.217(0.22)	4.74
$H_6(C_5)$	0.245(0.24)	0.177(0.14)	0.763(0.74)	3.56
$H_7(C_5)$	0.270(0.28)	0.067(0.09)	0.469(0.35)	3.56
$H_8(N)$	0.066(0.09)	0.128(0.16)	0.694(0.63)	4.10
$H_9(N)$	0.087	0.067	0.395	4.10

* The quantities in parentheses are from a differential electron-density synthesis.

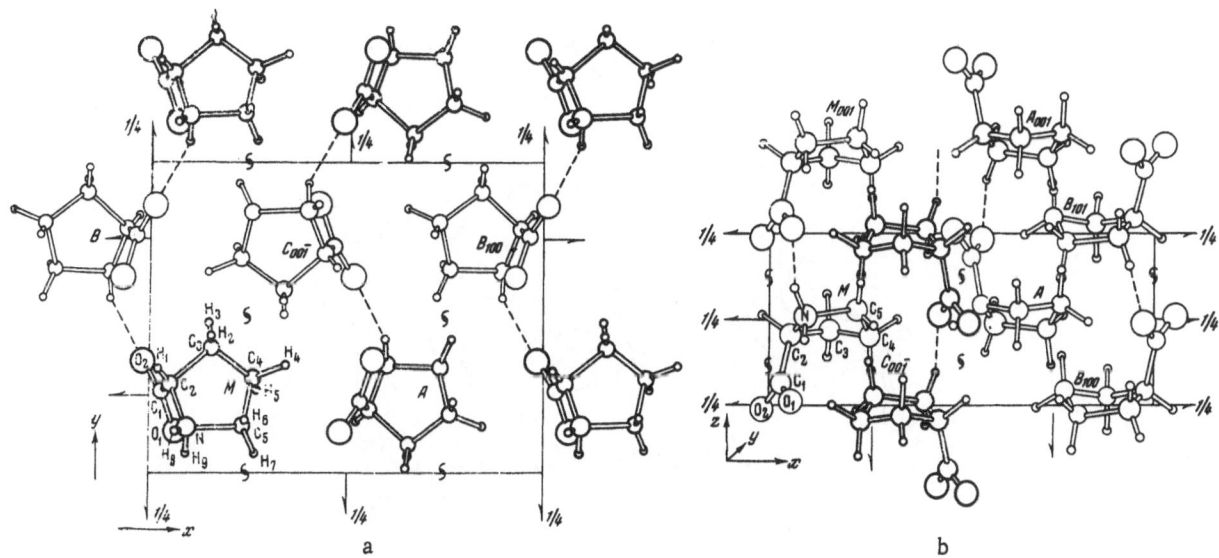

Fig. 83. Molecular array in the structure of L-proline: (a) view along c axis, (b) view along b axis.

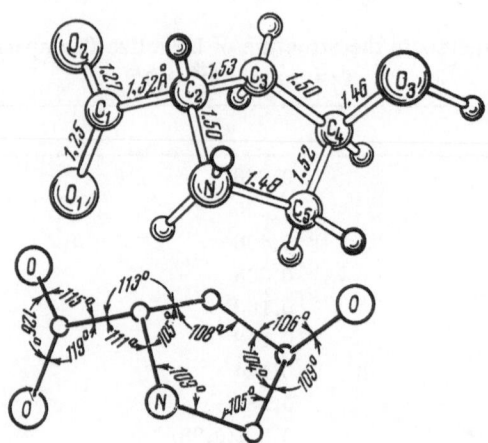

Fig. 84. Form of the molecule in the structure of L-hydroxyproline.

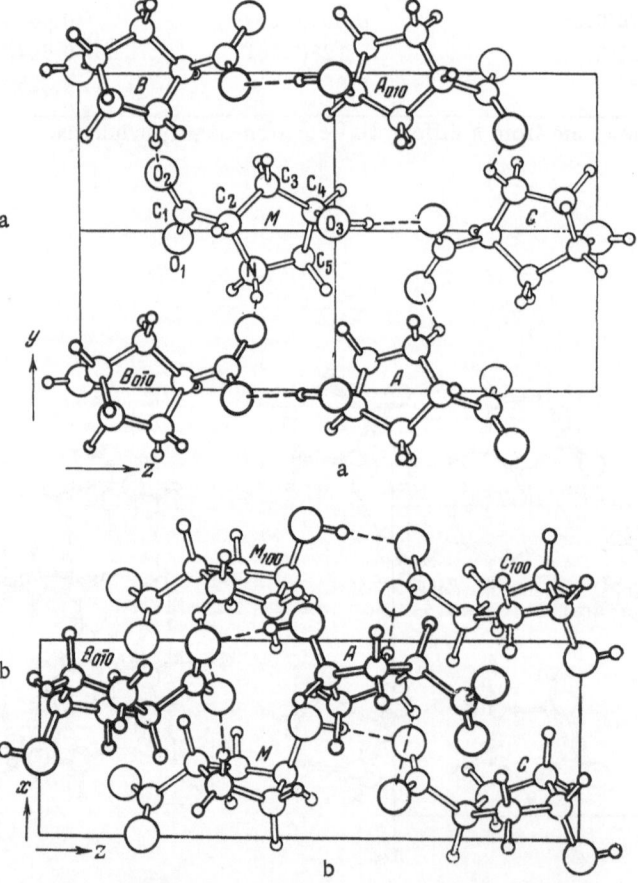

Fig. 85. Schematic representation of the structure of L-
hydroxyproline: (a) view along a axis, (b) view along b
axis.

TABLE 70. Atomic Coordinates in the Structure of L-Hydroxyproline

Atom	x	y	z	Atom	x	y	z
C_1	0,2062	0.5617	0.2036	$H_1(C_2)$	0.588	0,506	0,271
C_2	0,3763	0.5212	0.2938	$H_2(O_3)$	0.544	0.718	0.362
C_3	0.3647	0.6495	0.3711	$H_3^t(C_3)$	0.189	0.726	0.356
C_4	0,3290	0.5646	0.4637	$H_4(C_4)$	0.199	0.640	0.505
C_5	0.1866	0.4088	0.4375	$H_5(C_5)$	0.249	0.313	0.486
N	0.2774	0.3704	0.3407	$H_6^t(C_5)$	-0.030	0.418	0.435
O_1	0.0077	0.4746	0.1907	$H_7(N)$	0.434	0.290	0.347
O_2	0.2794	0.6850	0.1626	$H'_8(N)$	0.140	0.319	0.298
O_3	0.5979	0.5274	0.4974	$H_9(C_3)$	0,558	0,528	0,564

Hydroxyproline (4-Hydroxypyrrolidine-2-Carboxylic Acid)

This occurs in scleroproteins and keratins; it plays a major part in the secondary structure. It was isolated by Fischer in 1902 from acid hydrolyzates of gelatin, and shortly afterwards it was synthesized in the form of four stereoisomers (α- and γ-C asymmetric). The spatial structure of the naturally occurring isomer was examined first by chemical methods and then with x-rays [18-20, 127].

The first data on the crystal structure were reported by Zussman in 1951 [125, 126], but these were not very reliable, since the entire study was based on two-dimensional sets of intensities. A more careful study was performed in 1952 [127]. The crystals were grown from 95% ethanol by slow evaporation; space group $P2_12_12_1$, cell parameters a = 5.00 Å; b = 8.31 Å; c = 14.20 Å; ρ_{meas} = 1.474 g/cm^3; Z = 4.

The model was derived by interpreting the complete three-dimensional Patterson function, refinement being via the electron-density distribution, the coordinates of the hydrogen atoms being deduced from crystallographic considerations. This gave R(hk$l$) as 14.8%, with the distances to ~0.03 Å and the angles to ~1°. Table 70 gives the atomic coordinates.

Figure 84 shows the structure of a single molecule, in which C_4—O_3 (1.46 Å) is slightly more than the sum of the covalent radii (1.43 Å). The carboxyl group is almost symmetrical, which may imply a zwitterion form for the molecule. The pyrrolidine ring is appreciably distorted, C_4 (which bears the hydroxyl group) being 0.40 Å from the plane of the other four ring atoms, this forming a trans configuration with the carboxyl group. The carboxyl group and C_2 are coplanar, the nitrogen atom deviating 0.05 Å from the plane. In general, the relative configuration of the two asymmetric carbon atoms agrees with that previously deduced by chemical methods.

The molecules are linked into a three-dimensional net by hydrogen bonds; in two directions (Fig. 85) there are bonds between the nitrogen and carboxyl oxygen (N—H...O_2 = 2.69 and 3.17 Å), while in the third direction there are hydrogen bonds between the oxygen atoms of the carboxyl and hydroxyl groups (O_3—H...O_1 = 2.80 Å). There are also some fairly short C...O distances that may represent weak hydrogen bonds and may play a certain part in the molecular packing.

CHAPTER III

GENERAL RELATIONS OF MOLECULAR STRUCTURE TO PACKING IN CRYSTAL STRUCTURES FOR AMINO ACIDS

Table 71 summarizes the structure studies presented in the previous two chapters; it shows that the structures have been examined for all the principal amino acids except tryptophan, but not all the studies have been adequate. Moreover, demands have arisen for improved accuracy, because deductions about the stereochemistry of proteins are based on the results. This means that some of the structures require further refinement. All the same, the available evidence does enable one to draw some general conclusions about the structure of amino acids.

9. MOLECULAR STRUCTURES OF AMINO ACIDS

It is convenient here to consider separately the structures of the main groups: amino, carboxyl, and the radicals. Hahn [22] examined the evidence in this way for papers appearing up to 1957; the present discussion follows the same lines, with the addition of results appearing from 1957 to 1965 (Table 71).

Table 72 gives the data on the structures of amino groups, which in nearly all cases take the NH_3^+ form. The exception is arginine, where the unusual $-NH_2$ form occurs; but in this case there is a tetrahedral environment for the nitrogen atom via the formation of hydrogen bonds. Unfortunately, the accuracy of location of the hydrogen atoms by x-ray methods is inadequate to give any conclusions about the $N-H$ bond lengths or the detailed structure of the group.

The $C-N$ bond usually lies out of the plane of the carboxyl group, the deviation varying widely (Table 72).

The mean $C-N$ bond length for amino acids is 1.49_3 Å, which is greater than the standard for $C-N$ single bonds (1.47_5 Å [22]), which is usually ascribed to displacement of the nitrogen atom by tetrahedrally disposed hydrogen bonds and electrostatic forces between NH_3^+ and the negatively charged groups of adjacent molecules [22].

Tables 73 and 74 give the structural data for carboxyl groups. Hydrogen-atom localization for some structures has shown that the pure amino acids and their hydrates contain the group in the form COO⁻, whereas most of the hydrobromides and hydrochlorides contain the neutral COOII.

Following Pauling, we represent the structure of the carboxyl group as two principal resonant forms [128]:

$$R-C^I \begin{array}{c} O^I \\ \\ O^{II-} \end{array} \quad (I) \quad \text{and} \quad R-C^I \begin{array}{c} O^{I-} \\ \\ O^{II} \end{array} \quad (II)$$

Taking 1.425 and 1.205 Å as the standard lengths for $C-O^-$ and $C-O$ bonds, and 125.25° as the standard angle between them, we get the bond orders given in columns 4 and 7 of Table 73 [22] by reference to the curves relating bond length and valence angle respectively to bond order. This table shows that some amino acids have carboxyl groups nearly symmetrical as regards $C-O$ bonds, whereas others show a predominance of one

TABLE 71. X-Ray Data on Amino Acid Structures

Compound	Intensities used (3M three-dimensional, 2M two-dimensional) and space group	Accuracy ($\mathring{A}$) of distances
I. Aliphatic Amino Acids		
1. Monoaminomonocarboxylic acids		
α-glycine [24-30]	3M $P2_1/n$	0.005
β-glycine [31-33]	2M $P2_1$	0.015
γ-glycine [34-36]	3M $P3_1$	0.011
$Ni(gly)_2 \cdot 2H_2O$ [37]	2M $P2_1/c$	
$Cu(gly)_2 \cdot H_2O$ [38-41]	3M $P2_12_12_1$	0.010
Fe(II) sulfate-glycinate pentahydrate [42]	2M P1	
diglycine $\cdot$ HBr [43]	2M $P2_12_12_1$	0.03
diglycine $\cdot$ HCl [44]	2M $P2_12_12_1$	0.010
3gly $\cdot$ H_2SO_4 [45-46]	3M $P2_1$	
DL-alanine [51, 52]	3M Pna	0.010
L-alanine [53, 54]	3M $P2_12_12_1$	0.004
$Cu \cdot (\beta\text{-ala})_2 \cdot 6H_2O$ [55]	2M $P2_1/c$	
$Ni \cdot (\beta\text{-ala})_2 \cdot 2H_2O$ [56]	2M $P\bar{1}$	0.04
β-alanine [57]	3M Pbca	0.010
L-valine $\cdot$ HBr [58]	2M $P2_1$	
L-valine $\cdot$ HCl [59]	3M $P2_1$	0.018
L-valine $\cdot$ HCl $\cdot$ H_2O [60]	3M $P2_12_12_1$	
DL-valine [61]	P1 or $P\bar{1}$, cell parameters	
L-valine [62]	$P2_1$, cell parameters	
D-leucine [63]	$P2_122_1$, cell parameters	
DL-leucine [61]	P1, cell parameters	
L-leucine $\cdot$ HBr [64]	3M $P2_12_12_1$	0.06
D-isoleucine $\cdot$ HCl $\cdot$ H_2O [65]	2M $P2_12_12_1$	0.05
D-isoleucine $\cdot$ HBr $\cdot$ H_2O [65]	2M $P2_12_12_1$	0.05
DL-norleucine [66]	2M $P2_1/a$	0.04
2. Hydroxyaminomonocarboxylic acids		
DL-serine [67, 68]	3M $P2_1/a$	0.006
L-phosphoserine [69]	3M $P2_12_12_1$	0.02
L-threonine [70]	3M $P2_12_12_1$	0.010
3. Monoaminodicarboxylic acids		
DL-aspartic acid [61, 71]	2M I2/a	
Zn(L-asp) $\cdot$ $3H_2O$ [72]	3M $P2_12_12_1$	
glutamic acid HCl [73, 74]	2M $P2_12_12_1$	0.03
L-glutamic acid [75]	2M $P2_12_12_1$	0.05
$Cu \cdot glu \cdot 2H_2O$ [76]	3M $P2_12_12_1$	0.008
$Zn \cdot glu \cdot 2H_2O$ [77]	3M $P2_12_12_1$	0.008

TABLE 71 (continued)

Compound	Intensities used (3M three-dimensional, 2M two-dimensional) and space group	Accuracy (Å) of distances
5. Diaminomonocarboxylic acids		
L-arginine · $2H_2O$ [83]	3M $P2_12_12_1$	0.010
L-arginine · HBr · H_2O [84]	3M $P2_1$	0.03
L-arginine · HCl · H_2O [85]	3M $P2_1$	0.03
L-arginine · HCl [86]	3M $P2_1$	0.015
L-lysine · HCl · $2H_2O$ [86-90]	3M $P2_1$	0.015
6. Sulfur amino acids		
S-methyl-L-cysteine sulfoxide [92]	3M $P2_12_12_1$	0.02
L-cystine [93-95]	3M $P6_122$	0.007
L-cystine · 2HCl [96-98]	2M C2	
L-cystine · 2HBr [99-101]	2M $P2_122_1$	0.03
α-DL-methionine [103, 104]	2M $P2_1/a$	0.05
β-DL-methionine [103, 104]	2M $I2/a$	0.05
II. Aromatic and Heterocyclic Amino Acids		
1. Aromatic acids		
L-phenylalanine · HCl [106-108]	3M $P2_12_12_1$	0.02
L-tyrosine · HBr [109-111]	2M $P2_1$	0.05
L-tyrosine · HCl [109, 110, 112]	2M $P2_1$	0.03
2. Heterocyclic acids		
DL-tryptophan [61]	P1 or P1, cell parameters	
histidine · HCl · H_2O [114-116]	3M $P2_12_12_1$	0.010
Zn · (L-his)$_2$·$2H_2O$ [117]	3M $P4_12_12$	0.02
Zn · (DL-his)$_2$ · $5H_2O$ [118]	3M C2/c	0.02
L-proline [119, 122-124]	3M $P2_12_12_1$	0.02
Cu · DL · proline · $2H_2O$ [120]	2M $P2_1/n$	0.04
L-hydroxyproline [125-127]	3M $P2_12_12_1$	0.010

or other resonant form. The extent of this may sometimes be related to the distribution of the hydrogen bonds with respect to the two oxygen atoms. Inclusion in a strong X—H...O bond reduces the order of the C—O bond. For example, serine and threonine have equal hydrogen-bond energies for the two oxygen atoms, and hence complete resonance, whereas DL-alanine has a pronounced difference between these energies, which implies stabilization of one of the resonant forms.

The following are the resonant forms for un-ionized COOH groups:

$$\text{R—}C^I\underset{O^{II}H}{\overset{O^I}{{<}}} \quad (III) \quad \text{and} \quad \text{R—}C^I\underset{O^{II}H^+}{\overset{O^{I-}}{{<}}} \quad (IV)$$

However, calculations of bond order [22] reveal only weak resonance, form IV contributing only 15-20%

TABLE 72. Structural Data for Amino Groups

Compound	$C^{II}-NH_3^+$, Å	Valence angle $C^{I}-C^{II}-NH_3^+$, deg	Accuracy Bond-lengths, Å	Angles, deg	Deviation of $C^{II}-NH_3^+$ bond from plane of carboxyl group A	deg
α-glycine	1.474	111.8	0.005	0.5	0.436	18.6
β-glycine	1.484	110.8	0.015		0.583	24.8
γ-glycine	1.491	110.8	0.011		0.309	12.8
diglycine$_I$·HCl	1.521	109.4	0.010	0.7	0.330	12
diglycine$_{II}$·HCl	1.528	112.0	0.010	0.7	0.040	1
Cu·diglycine$_I$·H$_2$O	1.473	112.6	0.010-0.016		0.103	
Cu·diglycine$_{II}$·H$_2$O	1.484	111.3	0.010-0.016		0.162	
3 glycine$_I$·H$_2$SO$_4$	1.455	109.3			coplanar	
3 glycine$_{II}$·H$_2$SO$_4$	1.535	107.1			0.269	10.1
3 glycine$_{III}$·H$_2$SO$_4$	1.467	110.2			coplanar	
alanine	1.496	108.3	0.010	1	0.390	15
L-alanine	1.491	110.3	0.004		0.450	
valine·HCl	1.49	105.0	0.018	1.5		
valine·HCl·H$_2$O	1.49	108.0				
serine	1.491	110.0	0.006	0.5	coplanar	
phosphoserine	1.468		0.020		coplanar	
threonine	1.490	110.4	0.007	0.5	0.590	23
Zn·asp·3H$_2$O	1.490	106.0				
Cu·glu·2H$_2$O	1.486	110.1	0.008			
Zn·glu·2H$_2$O	1.475	110.7	0.008	0.4	coplanar	
asparagine·H$_2$O	1.500					
glutamine	1.51	111	0.02	1	1.18	51
arginine·2H$_2$O	1.480	110.9	0.01		0.280	
arginine$_I$·HCl	1.479	109.3	0.015	0.9		
arginine$_{II}$·HCl	1.502	108.1				
lysine·HCl·2H$_2$O	1.484	109.7	0.015		0.446	
cystine	1.511	108.5	0.007		0.105	4
S-methyl-L-cysteine sulfoxide	1.520	111.0	0.020	1-2	0.185	7
phenylalanine·HCl	1.480	106.0	0.020		0.058	2
histidine·HCl·H$_2$O	1.495	109.4	0.010			
Zn(his)$_2$·2H$_2$O	1.500	110	0.020	2	0.256	
Zn(his)$_2$·5H$_2$O	1.473	111.0	0.020	1.3	coplanar	
proline	1.530	107.4	0.020		0.23	
hydroxyproline	1.503	110.8	0.010	0.5	0.050	2
mean	1.493					
diglycine$_I$·HBr	1.52	111	0.03	3	0.49	19
diglycine$_{II}$·HBr	1.56	113	0.03	3	0.02	1
isoleucine·HCl	1.43	115	0.05	3	0.05	2
isoleucine·HBr	1.45	108	0.05	3	0.05	
α-norleucine	1.50	109	0.04	5		
glutamic acid	1.45	109	0.05	2	0.99	43
glutamic acid·HCl	1.52	111	0.03	1	0.44	18

TABLE 72 (continued)

Compound	$C^{II}-$ NH_3^+, Å	Valence angle $C^{I}-II-$ NH_3^+, deg	Accuracy		Deviation of $C^{II}-$ NH_3^+ bond from plane of carboxyl group	
			Bond-lengths, Å	Angles, deg	Å	deg
arginine$_I \cdot$ HCl $\cdot$ H$_2$O	1.52	108	0.03	2		
arginine$_{II} \cdot$ HCl $\cdot$ H$_2$O	1.51	107				
arginine$_I \cdot$ HBr $\cdot$ H$_2$O	1.49	110	0.03	2		
arginine$_{II} \cdot$ HBr $\cdot$ H$_2$O	1.45	110				
cystine $\cdot$ 2HCl	1.48	110				
cystine $\cdot$ 2HBr	1.49	108	0.03			
β-methionine	1.52	112	0.04	5	0.70	27
α-methionine	1.50	110	0.04	5	0.70	27
tyrosine $\cdot$ HCl	1.48	108	0.03		0.71	
tyrosine $\cdot$ HBr	1.49	109	0.05		0.62	
Cu $\cdot$ proline $\cdot$ 2H$_2$O	1.52	108	0.04	5	0.21	
Mean	1.49					

(Table 74). The deviation from pure single or double bonds in the carboxyl groups may be explained by analogy with the carboxyl ions (hydrogen bonds); a strong $O-H...X-Y$ bond tends to disrupt the $O-H$ bond and allows the $C-O$ bond to acquire some double-bond character, whereas the $X-Y$ bond may lose some double-bond character.

The bond orders of Table 74 are derived from the relation of bond order to length taking 1.395 and 1.185 Å as the standards for single and double $C-O$ bonds. These deviate somewhat from the standards used for carboxyl ions on account of the effect of polarity on bond length [22].

As regards the radicals R, we consider mainly the hydrocarbon chains; Table 75 illustrates the tendency to shortening and sometimes to alternation of $C-C$ bond lengths. The mean $C-C$ bond length is 1.52_3 Å, which is 0.02_1 Å less than the standard of 1.54_5 Å [22] for a single $C-C$ bond. The $C-C-C$ valence angles are simultaneously somewhat greater than the standard 109°, which shows that the $C-C$ bonds have a slight double-bond character, which is due to hyperconjugation of the $C-C$ bonds with the double $C=O$ bonds [22].

Molecules with long hydrocarbon radicals often show the effect known as rotational isomerism; groups linked by a multiple bond do not allow rotation around that bond, whereas rotation around single bonds is possible. Completely free rotation would allow a molecule with single bonds to exist as an infinite number of rotational isomers, whereas in fact there are always energy barriers, which produce the result that there is usually only one rotational isomer for a given single bond [4].

The various possible configurations of the molecule are indicated by the molecular structures found in the various modifications of the amino acids and their derivatives. The structural data give some evidence on the rotational isomers for methionine, isoleucine, cystine, and phenylalanine, which is of value in structure research on proteins. Attempts are being made to derive general configuration laws for the radicals R. It has been found that the γ-carbon and NH$_2$ nitrogen in all known structures have one of three possible positions with respect to the $C_\alpha - C_\beta$ bond: the trans position and the 60 and 300° gauche positions [87].

10. PACKING OF AMINO ACID MOLECULES IN CRYSTALS AND HYDROGEN BONDS

These topics are of particular interest, since the acids resemble most biopolymers in that the hydrogen bonds play a decisive role in the structure.

TABLE 73. Structural Data for Ionized Carboxyl Groups C^{II}–$C^{I}\langle O$–$O\rangle^{II}$–

Compound	C^{I}–C^{II} bond length (obs.), Å	O^{I}–C^{I}–O^{II} angle, deg	C^{I}–O^{I} and C^{I}–O^{II} bond lengths (obs.), Å	% double-bond character and theoretical length with standard of C–O=1.425 Å; C=O=1.205 Å (%)	Valence angles $\angle O^{I}$–C^{I}–C^{II} $\angle O^{II}$–C^{I}–C^{II} (obs.), deg	% double-bond character and theoretical valence angles $\angle O^{I}C^{I}C^{II}$ $\angle O^{II}C^{I}C^{II}$ (%)	% double-bond character and theoretical valence angles $\angle O^{I}C^{I}C^{II}$ $\angle O^{II}C^{I}C^{II}$ (deg)	Accuracy of observed values (Å)	Accuracy of observed values (deg)	Hydrogen-bond energy (kcal/mole) for oxygen atom
α-glycine	1.523	125.5	1.261 / 1.265	52 / 48	117.4 / 117.1	51 / 49	117.5 / 117.2	0.005	0.5	6.5 / 4.0
β-glycine	1.521	126.2	1.233 / 1.257		117.8 / 115.9			0.015		
γ-glycine	1.527	125.4	1.254 / 1.237		118.3 / 116.2			0.011		
diglycine$_{II}$·HCl	1.485	124.4	1.294 / 1.254	38 / 62	116.1 / 119.5	39 / 61	115.6 / 119.1	0.010	0.7	8 / 9
diglycine$_{II}$·HBr	1.50	128	1.24 / 1.22		119 / 114			0.030	3.0	
Cu·diglycine$_{I}$·2H$_2$O	1.498	124.3	1.275 / 1.226		117.4 / 118.3			0.016		
Cu·diglycine$_{II}$·2H$_2$O	1.541	122.8	1.291 / 1.243		117.5 / 119.7			0.016		
3 glycine$_{II}$·H$_2$SO$_4$	1.554	125.4	1.230 / 1.294		121.5 / 113.1					
alanine	1.536	125.4	1.273 / 1.211	32 / 68	113.2 / 121.3	24 / 76	113.3 / 121.4	0.010		7.5 / 4.5
L-alanine	1.525	125.6	1.256 / 1.247		116.1 / 118.3			0.004		
α-norleucine	1.49	122	1.26 / 1.20		120 / 118			0.04		
serine	1.528	125.3	1.261 / 1.268	52 / 48	117.2 / 117.4	49 / 51	117.2 / 117.5	0.006	0.5	7 / 7
threonine	1.517	126.9	1.253 / 1.236	45 / 55	116.1 / 117.0	47 / 53	116.9 / 117.8	0.007	0.5	7.5 / 6.5
Zn·asp·3H$_2$O	1.54	124	1.24 / 1.27		116 / 117	%	deg	Å	deg	
glutamic acid	1.55	127	1.27 / 1.24	45 / 55	115 / 117	45 / 55	116.5 / 118	0.05	2.0	9 / 7
Cu·glu·2H$_2$O	1.516	124.3	1.284		117.5			0.008		

Compound											
Cu · glu · $2H_2O$	1.509	120.6	1.242 1.296			118.2 116.8			0.008		
Zn · glu · $2H_2O$	1.537	125.2	1.242 1.268			122.6 115.1			0.008		
Zn · glu · $2H_2O$	1.523	121.0	1.245 1.288 1.246			119.7 116.0 123.0			0.008		
asparagine · H_2O	1.54		1.27 1.34								
glutamine	1.52	128	1.27 1.22	35 65	1.29 1.24	116 116	50 50	117.5 117.5	0.02	1.0	3.5 8.5
arginine · $2H_2O$	1.547	125.6	1.259 1.249			115.2 119.1			0.010		
lysine · HCl · $2H_2O$	1.529	125.5	1.250 1.246			116.8 117.7			0.015		
cystine	1.543	126.8	1.250 1.238			117.9 115.3			0.007		
α–methionine	1.47	121	1.28 1.21			120 119			0.05		
β–methionine	1.52	122	1.27 1.21			118 120			0.05		
histidine · HCl · H_2O	1.530	125.8	1.240 1.265			120.0 114.2			0.010		
Zn(his)$_2$ · $2H_2O$	1.53	123	1.20 1.25			118 119			0.02	2.0	
Zn(his)$_2$ · $5H_2O$	1.519	123	1.243 1.260			119 117			0.02	1.3	
proline	1.52	120	1.28 1.26			119 121			0.02		
Cu · proline · $2H_2O$	1.50	122	1.24 1.24			120 118			0.04	4.0	
hydroxyproline	1.516	126.1	1.269 1.254	45 55	1.269 1.252	115.4 118.5	40 60	115.7 118.9	0.010	0.5	6.5 4

TABLE 74. Structural Data for Un-ionized Carboxyl Groups C^{II}–$C^IO^IO^{II}H$

Compound	C^I–C^{II} bond length (obs.), Å	O^I–C^I–O^{II} angle, deg	C^I–O^I and C^I–O^{II} bond lengths (obs.), Å	% double-bond character and theoretical length with standard of C–O=1.425 Å; C=O=1.205 Å %	Å	Valence angles ∠O^I–C^I–C^{II} ∠O^{II}–C^I–C^{II} (obs.), deg	% double-bond character and theoretical valence angles ∠$O^IC^IC^{II}$ ∠$O^{II}C^IC^{II}$ %	deg	Accuracy of observed values Å	deg	Hydrogen-bond energy (kcal/mole) for oxygen atom
diglycine·HCl	1.480	124.1	1.320 / 1.235	20 / 80	1.313 / 1.219	113.0 / 122.9	18 / 82	112.3 / 122.4	0.010	0.7	9
diglycine·HBr	1.52	128	1.29 / 1.25			121 / 112			0.03	3	1
3 glycine$_I$·H$_2$SO$_4$	1.547	125.9	1.289 / 1.230			115.3 / 115.1					
3 glycine$_{II}$·H$_2$SO$_4$	1.513	122.9	1.314 / 1.251			111.3 / 119.2					
valine·HCl	1.50	121	1.35 / 1.23			112 / 127			0.02	1.5	
valine·HCl·H$_2$O	1.49	116	1.30 / 1.27			121 / 123					
isoleucine·HCl	1.51	124	1.27 / 1.27			119 / 117			0.05		
isoleucine·HBr	1.51	125	1.33 / 1.22			114 / 121			0.05		
phosphoserine	1.541		1.321 / 1.201						0.02		
glutamic acid	1.57	130	1.24 / 1.19	95	1.19	108 / 122	5 / 95	110 / 124	0.02	2	9
glutamic acid$_I$·HCl	1.51	124	1.31 / 1.21	20 / 80	1.30 / 1.20	114 / 123	20 / 80	113 / 122	0.03	1	2.5
glutamic acid$_{II}$·HCl	1.54	125	1.32 / 1.20	15 / 85	1.32 / 1.20	112 / 124	15 / 85	112 / 123	0.03	1	9
arginine$_I$·HBr·H$_2$O	1.57	127.9	1.28 / 1.27			113.5 / 118.5			0.03	1.9	3
arginine$_{II}$·HBr·H$_2$O	1.55	122.9	1.26 / 1.20			117.7 / 119.0			0.03	1.9	9
arginine$_I$·HCl·H$_2$O	1.52	126.8	1.29 / 1.28			118.5 / 114.5			0.03	1.8	
arginine$_{II}$·HCl·H$_2$O	1.52	123.5	1.31 / 1.23			120.8 / 115.6			0.03	1.8	

arginine I · HCl	1.539	125.6	1.262	115.1	0.015	0.9
			1.256	118.9		
arginine II · HCl	1.579	126.0	1.253	117.6	0.015	0.9
			1.251	116.4		
cystine · 2HCl	1.474	119.1	1.307	118.1		
			1.238	122.7		
cystine · 2HBr	1.506	123.8	1.269	114.2	0.03	
			1.215	122.0		
tyrosine · HCl	1.510	121	1.270	122	0.03	
			1.265	118		
tyrosine · HBr	1.55	126	1.29	116	0.05	
			1.26	116		
phenylalanine · HCl	1.50	124	1.34	109	0.02	
			1.17	127		
histidine · HCl	1.530	125.8	1.265	114.2	0.010	
			1.240	120.0		

TABLE 75. Lengths of C—C Bonds and C—C—C Valence Angles in Aliphatic Parts of Molecules

Compound	Distance C^I–C^{II}, Å	Valence angles C^I–C^{II}–C^{III}, deg	Distance C^{II}–C^{III}, Å	Valence angles C^{II}–C^{III}–C^{IV}, deg	Distance C^{III}–C^{IV}, Å	Valence angles C^{III}–C^{IV}–C^V, deg	Distance C^{IV}–C^V, Å	Valence angles C^{IV}–C^V–C^{VI}, deg	Distance C^V–C^{VI}, Å
α-glycine	1.523								
β-glycine	1.521								
γ-glycine	1.527								
diglycineI · HCl	1.480								
diglycineII · HCl	1.485								
Cu · diglycineI · H₂O	1.498								
Cu · diglycineII · H₂O	1.541								
3 glycineI · H₂SO₄	1.547								
3 glycineII · H₂SO₄	1.513								
3 glycineIII · H₂SO₄	1.554								
alanine	1.536	111	1.513						
L-alanine	1.525	111.5	1.525						
valine · HCl	1.50	112	1.52	115	1.55	112	1.50	112	
valine · HCl · H₂O	1.49	110	1.54	111	1.58	110	1.49	112	
serine	1.528	110	1.513						
phosphoserine	1.541		1.526						
threonine	1.517	113	1.542	113	1.505				
Zn · asp · 3H₂O	1.54	112	1.58	116	1.45				
Cu · glu · 2H₂O	1.516	110.0	1.530	115.4	1.510	112.4	1.509		
Zn · glu · 2H₂O	1.537	110.2	1.517	114.0	1.517	113.5	1.523		
asparagine · H₂O	1.54	114	1.51		1.53				
glutamine	1.52		1.50	113	1.47	115	1.54		
arginine · 2H₂O	1.547	108.4	1.542	114.4	1.540	110.1	1.517		
arginineI · HCl	1.539	112.5	1.534	113.2	1.555	109.7	1.543		
arginineII · HCl	1.579	110.2	1.520	114.2	1.498	110.0	1.516		
lysine · HCl · 2H₂O	1.529	109.8	1.524	114.6	1.518	111.0	1.526	111.5	1.521
cystine	1.543	114.2	1.509						
S-methyl-L-cysteine sulfoxide	1.57	110	1.49						
phenylalanine · HCl	1.50	116	1.55	113	1.57			Benzene ring	

histidine · HCl · H₂O	1.530	113.3	1.527	114.9	1.508		Ring	
Zn · (his)₂ · 5H₂O	1.519	110	1.539	114	1.486		Ring	
Zn · (his)₂ · 2H₂O	1.53	111	1.52	113	1.48		Ring	
proline	1.52	112	1.52	101	1.54	102	1.52	
hydroxyproline	1.516	113.3	1.532	107.6	1.503	104	1.524	1.48

Mean C—C bond length 1.52_3 Å, mean C—C—C angle $111_{96}°$

diglycineI · HBr	1.52							
diglycineII · HBr	1.50							
isoleucine · HCl	1.51	104	1.55	109	1.53	119	1.52	
isoleucine · HBr	1.51	113	1.54	112	1.56	113	1.54	
α-norleucine	1.49	112	1.58	114	1.51	110	1.57	120
glutamic acid	1.55	110	1.56	117	1.54	114	1.57	
glutamic acid · HCl	1.51	109	1.55	115	1.51	109	1.54	
arginineI · HBr · H₂O	1.57	110.2	1.53	110.9	1.58	106	1.55	
arginineII · HBr · H₂O	1.55	107.1	1.57	109.5	1.61	106.5	1.52	
arginineI · HCl · H₂O	1.52	106.7	1.59	106.3	1.57	103.9	1.58	
arginineII · HCl · H₂O	1.52	105.7	1.58	108.9	1.56	105.6	1.58	
cystine · 2HCl	1.474	111.5	1.561					
cystine · 2HBr	1.509	113.9	1.506					
α-methionine	1.47	111	1.55	111	1.51			
β-methionine	1.52	108	1.58	113	1.54			
tyrosine · HCl	1.510	110	1.530	113	1.522		Benzene ring	
tyrosine · HBr	1.55	111	1.47	116	1.57		Benzene ring	
Cu · proline · 2H₂O	1.50	112	1.52	97	1.50	109	1.52	

Mean C—C bond length 1.53 Å, mean C—C—C angle $110_3°$

TABLE 76. Bonds between Molecules in Amino Acids

Compound	Groups forming hydrogen bonds	H-bond system and mode of packing
Monoaminomonocarboxylic Acids		
α-glycine	COO, NH_3	two-dimensional network, molecules in double layers linked by van der Waals forces
β-glycine	COO, NH_3	three-dimensional network forming a framework
γ-glycine	COO, NH_3	the same
$Cu \cdot$ diglycine $\cdot H_2O$	COO, NH_2, H_2O	the same
diglycine $\cdot$ HCl (HBr)	$COOH$, COO, NH_2, NH_3, ions Cl^- (Br^-)	the same
3 glycine$_{II} \cdot H_2SO_4$	$COOH$, COO, NH_2, NH_3, sulfate ions	the same
DL-alanine	COO, NH_3	the same
L-alanine	COO, NH_3	the same
L-valine $\cdot$ HCl $\cdot H_2O$	$COOH$, NH_3, ions Cl^- (Br^-)	two-dimensional network
L-valine $\cdot$ HCl $\cdot H_2O$	$COOH$, NH_3, H_2O, ions Cl^-	the same
L-leucine $\cdot$ HBr	$COOH$, NH_3, ions Br^-	the same
D-isoleucine $\cdot$ HCl (HBr) $\cdot H_2O$	$COOH$, NH_3, ions Cl^- (Br^-), H_2O	the same
Hydroxylmonoaminocarboxylic Acid		
DL-serine	COO, NH_3, OH	the same
L-threonine	COO, NH_3, OH	three-dimensional network
Monoaminodicarboxylic Acids		
DL-aspartic acid	$COOH$, COO, NH_3	three-dimensional network
glutamic acid $\cdot$ HCl	α-$COOH$, γ-$COOH$, ions Cl^-	the same
L-glutamic acid	COO, $COOH$, NH_3	the same
L-alanine	COO, NH_3	the same
$Cu \cdot$ glu $\cdot 2H_2O$	NH, COO, $2H_2O$	the same
$Zn \cdot$ glu $\cdot 2H_2O$	NH, COO, $2H_2O$	the same
Amides of Monoaminodicarboxylic Acids		
L-asparagine $\cdot H_2O$	COO, NH_3, O, and NH_2 amide group, H_2O	three-dimensional network
L-glutamine	COO, NH_3, O, and NH_2 amide group	the same
Diaminomonocarboxylic Acid		
L-arginine $\cdot 2H_2O$	COO, N of amino group, $(NH_2)_2$, and NH guanidine group, $2H_2O$	three-dimensional network

TABLE 76 (continued)

Compound	Groups forming hydrogen bonds	H-bond system and mode of packing
L-lysine·HCl·2H$_2$O	COO, α-NH$_3$, ε-NH$_3$, ions Cl$^-$, 2H$_2$O	three-dimensional network
Sulfur Amino Acids		
L-cystine	(COO, NH$_3$)$_2$	three-dimensional network,
L-cystine·2HCl (HBr)	(COOH; NH$_3$, ions Cl$^-$ [Br$^-$])$_2$	the same
α (β)-DL-methionine	COO, NH$_3$	two-dimensional network, layered structure
Aromatic Amino Acids		
L-phenylalanine·HCl	COOH, NH$_3$, ions Cl$^-$	two-dimensional network, layered structure
L-tyrosine·HCl (HBr)	COOH, NH$_3$, OH, ions Cl$^-$ (Br$^-$)	three-dimensional network
Heterocylic Amino Acids		
histidine·HCl·H$_2$O	COOH, NH$_3$, (NH)$_2$-imidazole ring Cl$^-$, H$_2$O	three-dimensional network
Zn·(DL-his)$_2$·5H$_2$O	COO, NH$_2$, NH-imidazole ring 5H$_2$O	the same
Zn·(L-his)$_2$·2H$_2$O	COO, NH$_2$, NH imidazole ring 2H$_2$O	the same
L-proline	COO, NH$_2$	two-dimensional network, layered structure
Cu·DL-proline·2H$_2$O	COO, NH, 2H$_2$O	the same
L-hydroxyproline	COO, NH$_2$, OH	three-dimensional network

The energy of a hydrogen bond (6-10 kcal/mole) is higher than that of van der Waals interaction (1-2 kcal/mole), so first all possible hydrogen bonds are formed and then the closest packing is produced.

Table 57 shows that all the structures, except for L-arginine·2H$_2$O and the copper and zinc salts of glutamic acid, have all relevant groups participating in the hydrogen bonds.

The monoaminomonocarboxylic and aromatic amino acids have no additional active groups, and here the hydrogen bonds are formed solely between the amino and carboxyl groups. If the molecule is small and compact, the hydrogen bonds run throughout the crystal and join the molecules into a three-dimensional framework, as in α- and γ-glycine and DL-alanine.

Glycine is of interest in that its very small size allows it to pack in the crystal in several ways, each with its own system of hydrogen bonds (see Figs. 10, 13, and 14 for the α-, β-, and γ-modifications of glycine). The molecule has almost the same structure in each of these.

If the molecule has an aromatic ring or a long aliphatic chain, we find a two-dimensional hydrogen-bond network. Such structures are layered and show good cleavage along certain planes. For instance, the principal structural elements are double layers in L-phenylalanine·HCl, D-isoleucine·HCl (or HBr), and α- or β-DL-methionine. Each double layer contains molecules joined together by hydrogen bonds, while

TABLE 77. Lengths (Å) of Y—H...Z Hydrogen Bonds and X—Y...Z Angles (Degrees) Found in Amino Acid Structures

Compound	N—H...O between NH₃⁺ and COO⁻ (or COOH)	N—H...Cl⁻ (Br⁻) between NH₃⁺ and Cl⁻ (Br⁻)	N—H...O between NH₂ and C—O (chains)	N—H...O between NH₃⁺ and H₂O	Between carboxyl groups (N—H...Z^I; Z^II between NH₃⁺ and Z^I; Z^II)	O—H...Cl⁻ (Br⁻)	O—H...O between H₂O and COO⁻	O—H...O between C—OH and COO⁻	O—H...O between carboxyl groups
α-glycine	2.85 Å (118°), 2.85 (116)				NH_3^+ ⟨ COO^- 2.95 Å				
γ-glycine	2.80 (111), 2.81 (91), 2.97 (124)				COO^- 3.07				
β-glycine	2.76, 2.83				NH_3^+ ⟨ COO^- 3.02				
diglycine·HCl	2.93 (102), 2.90 (114)	3.32 Å (96°), 3.13 (115)			COO^- 3.00, COO^- 2.98				2.57 Å
diglycine·HBr	2.94 (101), 2.90 (116)	3.22 (99), 3.37 (110), 3.37 (81), 3.33 (103)			COO^- 3.04 COOH, COO^- 3.04, NH_3^+ ⟨ COO^- 3.10 COOH				2.46
Cu·diglycine·H₂O	3.03, 3.09, 2.99						2.80 Å, 2.76		
3 glycine·H₂SO₄	2.717, 2.823								2.438
DL-alanine	2.88, 2.84, 2.80								
L-alanine	2.813, 2.853, 2.829								
L-valine·HCl		3.22, 3.38			N—H ⟨ Cl (3.20); O_1 (2.99)	3.03			
L-valine·HCl·H₂O		3.26		2.90, 2.87		3.01, 3.17	3.19		
D-isoleucine·HCl·H₂O		3.18		2.85 Å, 2.96		donor H₂O 3.24 Å, donor COOH 3.05, donor H₂O 3.07			
D-isoleucine·HBr·H₂O		3.35		2.82, 2.84		donor COOH 3.39, donor H₂O 3.30, donor H₂O 3.35			

Compound	Values
DL-norleucine	2.73 (109), 2.87 (108), 2.87 (105)
DL-serine	2.81 (121); NH₃⁺–HOC⁻ 2.79 Å (99°); 2.67 Å
phosphoserine	2.79, 2.83
L-threonine	2.90 (116), 2.80 (98); 2.66
L-glutamic acid	2.86 (102), 2.92 (103), 2.94 (113); 2.54
glutamic acid HCl	2.89 (88), 3.18, 3.18; donor COOH 3,06; 2.57
Cu · glu · 2H₂O	2.966; 2.781, 2.722, 2.730, 2.758
Zn · glu · 2H₂O	2.975; 2.84, 2.80
L-asparagine · H₂O	2.80, 2.81; 2.92, 3.02; NH₃...OC 2.85
L-glutamine	2.79 (98), 2.85 (117); 2.94 (145), 2.91; NH₃...OC 2.91 (125)
L-arginine · 2H₂O (amino group –NH₂, with 2(NH₂) and NH in guanidine group) ·	2.897 (135), 2.869 (116), 2.887 (121); 2.912; 2.739, 2.890
L-lysine · HCl · 2H₂O	2.79 (111), 2.79 (94); 3.22 (111), 3.17 (111); 2.81 (107), 3.35, 2.89; 2.80
L-cystine	2.79, 2.81, 2.86
L-cystine · 2HCl	3.08, 3.27, 3.25, 3.28, 3.42, 3.41
L-cystine · 2HBr	donor COOH 2.98
α-DL-methionine	2.92, 2.59, 2.80; donor COOH 3.17
β-DL-methionine	2.82, 2.80, 2.78

TABLE 77 (continued)

Compound	N–H...O between NH$_3^+$ and COO$^-$ (or COOH)	N–H...Cl$^-$ (Br$^-$) between NH$_3^+$ and Cl$^-$ (Br$^-$)	N–H...O between NH$_2$ and C–O (chains)	N–H...O between NH$_3^+$ and H$_2$O	Between carboxyl groups N–H...ZI, ZII between (NH$_3$)$^-$ and ZI, ZII	O–H...Cl$^-$ (Br$^-$)	O–H...O between H$_2$O and COO$^-$	O–H...O between C–OH and COO$^-$	O–H...O between carboxyl groups
L-tyrosine · HCl		3.31 (137) 3.32 (86) 2.79 (124)				donor C–OH 3.11		donor COOH acceptor C–OH 2.50	
L-tyrosine · HBr		3.46 (101) 3.46 (136) 3.50 (107)				donor C–OH 3.23		donor COOH acceptor C–OH 2.50	
L-phenylalanine · HCl		3.15 3.30			3.23 3.06	donor COOH 2.94			
histidine · HCl · H$_2$O		3.186 3.204	2.637 2.846	2.816 2.86					
Zn · (L-his)$_2$ · 2H$_2$O	2.90 2.88					donor H$_2$O 3.199	2.745 2.83 2.75		
Zn · (DL-his)$_2$ · 5H$_2$O	2.762 2.963			3.008			2.964 2.724		
L-proline (amino group NH$_2^+$)	2.71 2.69								
Cu · DL-proline · 2H$_2$O	2.86						2.94		
L-hydroxyproline (amino group NH$_2^+$)	2.69 3.17						3.00	2.80	

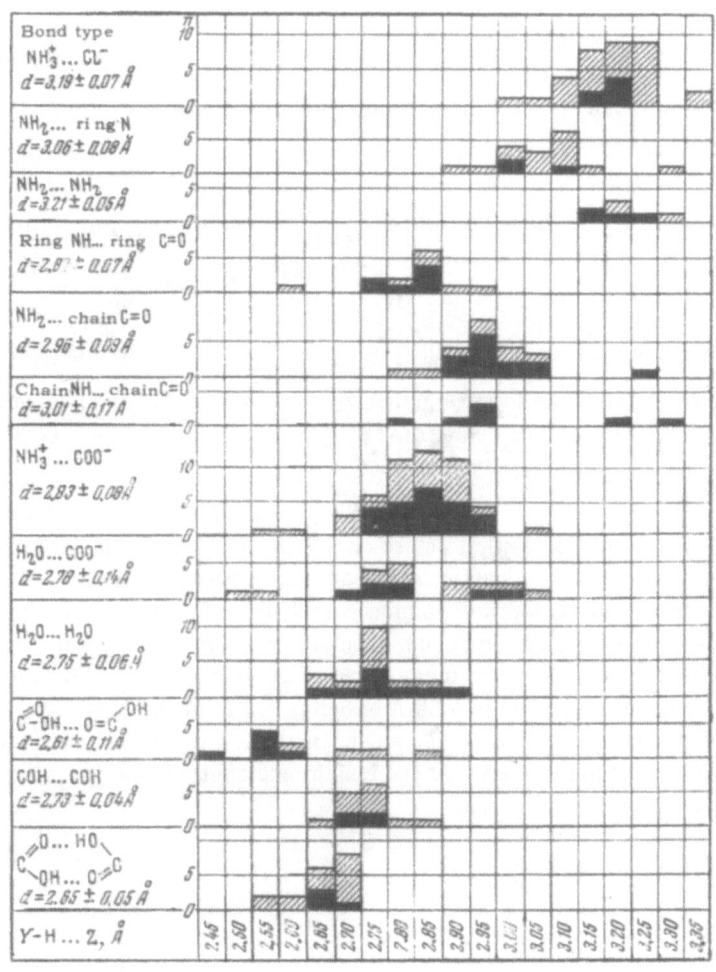

a

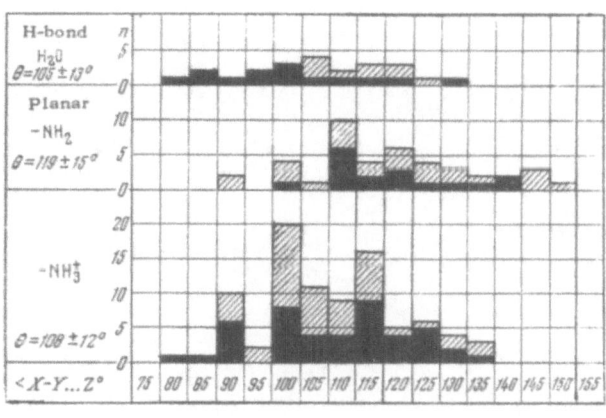

b

Fig. 86. Distribution of hydrogen bonds: (a) by length, (b) by angle.

between layers there are only weak van der Waals forces, which explains the cleavage. The van der Waals forces have most effect on the packing in layered structures, and polymorphism is then common.

If the molecule contains additional groups capable of forming hydrogen bonds, there are always three-dimensional hydrogen-bond networks, no matter what the size of the molecule (Table 76).

Hydrogen bonds cause the projections in one molecule (the H atoms of OH or NH) to approach the projections (O, N, etc., atoms) in another, so a center of symmetry or a twofold axis (simple or screw) may relate molecules of arbitrary shape [4].

The molecular asymmetry (existence of mirror-image forms) imposes a major restriction on the space groups open to amino acids, since the molecules have no symmetry elements of the second kind. Hence the optically active amino acids must crystallize mainly in groups with simple or screw twofold axes; in fact, the great majority of structures belong to groups $P2_12_12_1$ and $P2_1$ (Table 71), which are also the most probable from the viewpoint of the theory of close packing. The principles of the latter thus play a major part here.

A racemate crystal contains both forms together, so a twofold symmetry element must be present. Packing theory indicates that this element may be an inversion center or a glide plane [25], while Table 71 shows that these elements are present in the structures of racemates.

The types of hydrogen bond* are determined by the groups COO^-, $COOH$, NH_3+, CO, NH_2+, OH, together with the Cl^- and Br^- ions, the H_2O molecules, and so on. Table 77 gives the lengths and the corresponding $X-Y...Z$ angles (Y is the donor, Z the acceptor, and X an atom covalently linked to Y). Figure 86 shows the distributions of the lengths and angles for the commonest bond types [129]; the data derive from the electron or neutron diffraction for various classes of compound. The ordinates are the numbers of bonds in the interval $\Delta = 0.05$ Å or $\Delta' = 5°$, while the abscissa is the mean value of the interval. The black rectangles correspond to bonds whose lengths have been established to 0.05 Å or better, while the hatched parts represent ones of error limits greater than 0.05 Å. The mean lengths and angles (with their standard deviations) are also given. Comparison of Table 77 with the diagrams shows that the lengths and angles found in amino acids fall within the limits usually encountered in other compounds.

---

*H-bonds are usually classified by reference to the types of group involved. Each type of bond has a certain mean length and range of variation [129].

LITERATURE CITED

1. Transactions of the Fifth International Biochemical Congress: Biological Structures and Functions at the Molecular Level, Moscow, 1962.
2. M. F. Perutz. Proteins and Nucleic Acids, New York, American Elsevier, 1962.
3. Z. B. Low. In: The Proteins, edited by H. Neurath and K. Bailey, Vol. 2, New York, Academic Press, 1954.
4. B. K. Vainshtein. Diffraction of X-Rays by Chain Molecules, Moscow, Izd. Akad. Nauk SSSR, 1963.
5. S. E. Bresler. Introduction to Molecular Biology, Moscow-Leningrad, Izd. Akad. Nauk SSSR, 1963.
6. B. K. Vainshtein. Vestnik Akad. Nauk SSSR, 12:20 (1960).
7. N. S. Andreeva. Uspekhi sovremennoi biologii, 58:3 (1964).
8. J. S. Kendrew. Biofizika, 8:273 (1963).
9. M. H. F. Wilkins. Biofizika, 8:641 (1963).
10. C. C. F. Blake, D. F. Koenig, G. A. Mair, A. C. T. North, D. C. Phillips, and V. R. Sarma. Nature, 206:757 (1965).
11. The Principles of Molecular Biology: Virology and Immunology, Moscow, Izd. Nauka, 1964.
12. A. Rich and D. W. Green. Ann. Rev. Biochem., 30:93 (1961).
13. Current Problems in Biochemistry [Russian translation], Moscow, IIL, 1957.
14. R. E. Dickerson. The Proteins, Academic Press, New York, 1964.
15. D. C. Phillips. Adv. in Protein Crystallography (in press).
16. B. K. Vainshtein. X-Ray structure analysis of globular proteins, Uspekhi fiz. nauk, 88:527 (1966).
17. L. Pauling, R. B. Corey, and H. R. Branson. Proc. Nat. Acad. Sci. Wash., 37:205 (1951); Proc. Roy. Soc., B141:21 (1953).
18. A. Meister. Biochemistry of the Amino Acids, New York, Academic Press, 1965.
19. F. B. Straub. Biochemistry, Budapest, Hungarian Academy of Sciences, 1963.
20. N. K. Kochetkov, I. V. Torgov, and M. M. Botvinik. The Chemistry of Natural Compounds, Moscow, Izd. Akad. Nauk SSSR, 1961.
21. J. P. Greenstein and M. Winitz. Chemistry of the Amino Acids, New York, Wiley, 1961.
22. T. Hahn. Z. Krist., 109:438 (1957).
23. J. D. Bernal. Z. Krist., 78:363 (1931).
24. G. Albrecht and R. B. Corey. J. Am. Chem. Soc., 61:1087 (1939).
25. A. I. Kitaigorodskii. Organic Chemical Crystallography, New York, Consultants Bureau, 1961.
26. J. Hengstenberg and F. V. Lenel. Z. Krist., 77:424 (1931).
27. A. I. Kitaigorodskii. Acta Physiochim. URSS, 5:749 (1936).
28. R. B. Corey. Fortschr. Chem. Organ. Naturstoffe, 8:310 (1951).
29. R. E. Marsh. Acta Cryst., 10:814 (1957).
30. R. E. Marsh. Acta Cryst., 11:654 (1958).
31. C. J. Ksanda and G. Tunell. Am. J. Sci. A, 35:173 (1938).
32. Y. Iitaka. Acta Cryst., 6:663 (1953).
33. Y. Iitaka. Acta Cryst., 13:35 (1960).
34. Y. Iitaka. Proc. Japan Acad., 30:109 (1954).
35. Y. Iitaka. Acta Cryst., 11:225 (1958).

36.   Y. Iitaka. Acta Cryst., 14:1 (1961).

37.   A. J. Stosick. J. Am. Chem. Soc., 67:362 (1945).

38.   J. Okaya et al. Acta Cryst., 10:799 (1957).

39.   K. Tomita and I. Nitta. Bull. Chem. Soc. Japan, 34:286 (1961).

40.   A. V. Ablov, I. A. D'yakov, V. Ya. Ivanova, N. N. Proskina, and L. F. Chapurina. Neorgan. khimiya, 10:628 (1965).

41.   H. C. Freeman, M. R. Snow, I. Nitta, and K. Tomita. Acta Cryst., 17:1463 (1964).

42.   I. Lindquist and R. Rosenstein. Acta Chem. Scand., 14:1228 (1960).

43.   M. J. Buerger, E. Barney, and T. Hahn Z. Krist., 108:130 (1956).

44.   T. Hahn and M. J. Buerger. Z. Krist., 108:419 (1957).

45.   E. A. Wood and A. H. Holden. Acta Cryst., 10:145 (1957).

46.   S. Hoshino, Y. Ikaya, and R. Pepinsky. Phys. Rev., 115:323 (1959).

47.   R. Pepinsky, Y. Okaya, and F. Jona. Bull. Am. Phys. Soc., Ser. II, 2:220 (1957).

48.   J. W. Turley et al. Acta Cryst., 10:813 (1957).

49.   R. Pepinsky, V. Veda, and Y. Okaya. Phys. Rev., 110:1309 (1958).

50.   R. Pepinsky et al. Phys. Rev., 107:1538 (1957).

51.   H. A. Levy and R. B. Corey. J. Am. Chem. Soc., 63:2045 (1941).

52.   J. Donohue. J. Am. Chem. Soc., 72:949 (1950).

53.   H. J. Simpson, Jr., and R. E. Marsh. Acta Cryst., 20:550 (1966).

54.   J. D. Dunitz and R. R. Ryan. Acta Cryst., 21:617 (1966).

55.   K. Tomita. Bull. Chem. Soc. Japan, 34:297 (1961).

56.   P. Jose, L. M. Pant, and A. B. Biswas. Acta Cryst., 17:24 (1964).

57.   P. Jose and L. M. Pant. Acta Cryst., 18:806 (1965).

58.   R. Parthasarathy and R. Chandrasekaran (to be published).

59.   R. Parthasarathy and G. N. Ramachandran (to be published).

60.   S. Thyagaraja Rao (to be published).

61.   B. Dawson and A. McL. Mathieson. Acta Cryst., 4:475 (1951).

62.   M. Tsuboi, T. Takenishi, and Y. Iitaka. Bull. Chem. Soc. Japan, 32:305 (1959).

63.   K. Möller. Acta Chem. Scand. 3:1326 (1949).

64.   E. Subramanian (to be published).

65.   J. Trommel and J. M. Bijvoet. Acta Cryst., 7:703 (1954).

66.   A. McL. Mathieson. Acta Cryst., 6:399 (1953).

67.   G. Albrecht, G. W. Schnakenberg, M. S. Dunn, and J. D. McCullough. J. Phys. Chem., 47:24 (1943).

68.   D. P. Shoemaker, R. E. Barieau, J. Donohue, and Chia-Si Lu. Acta Cryst., 6:241 (1953).

69.   G. H. McCallum, J. M. Robertson, and G. A. Sim. Nature, 184:1863 (1959).

70.   D. P. Shoemaker, J. Donohue, V. Schomaker, and R. B. Corey. J. Am. Chem. Soc., 72:2328 (1950).

71.   V. Amirthalingam and G. N. Ramachandran. Current Sci. (India), 24:294 (1955).

72.   T. Doyne, R. Pepinsky, and T. Watanabe. Acta Cryst., 10:438 (1957).

73.   B. Dawson. Acta Cryst., 6:81 (1953).

74.   W. C. Quene and F. Jellinek. Acta Cryst., 12:439 (1959).

75.   S. Hirokawa. Acta Cryst., 8:637 (1955).

76.   C. M. Gramaccioli and R. E. Marsh. Acta Cryst., 21:594 (1966).

77.   C. M. Gramaccioli. Acta Cryst., 21:600 (1966).

78.   F. C. Steward and J. F. Thompson. Nature, 169:739 (1952).

79.   Y. Saito, O. Cano-Corona, and R. Pepinsky. Science, 121:435 (1955).

80.   R. A. Pasternak, L. Katz, and R. B. Corey. Acta Cryst., 7:225 (1954).

81.   G. Kartha and A. deVries. Nature, 192:862 (1961).

82.   W. Cochran and B. R. Penfold. Acta Cryst., 5:644 (1952).

83.   I. L. Karle and J. Karle. Acta Cryst., 17:835 (1964).

84.   S. K. Mazumdar and R. Srinivasan. Current Sci. (India), 33:573 (1964).

85.   S. K. Mazumdar and R. Srinivasan (to be published).

86. K. Vankatesan, S. K. Mazumdar, H. C. Mez, and J. Donohue (to be published).
87. G. N. Ramachandran, S. K. Mazumdar, K. Venkatesan, and A. V. Lakshminarayanan (to be published).
88. R. Srinivasan. Acta Cryst., 9:1039 (1956).
89. S. Raman. Z. Krist., 111:301 (1959).
90. D. A. Wright and R. E. Marsh. Acta Cryst., 15:54 (1962).
91. B. K. Vainshtein. Kristallografiya, 3:3 (1958).
92. R. Hine. Acta Cryst., 15:635 (1962).
93. L. K. Steinrauf and L. H. Jensen. Acta Cryst., 9:539 (1956).
94. B. M. Oughton and P. M. Harrison. Acta Cryst., 10:479 (1957).
95. B. M. Oughton and P. M. Harrison. Acta Cryst., 12:396 (1959).
96. A. F. Corsmit, A. Schuyff, and D. Feil. Proc. Konink. Ned. Akad. Wet., B59:470 (1956).
97. L. K. Steinrauf and L. H. Jensen. Acta Cryst., 10:814 (1957).
98. L. K. Steinrauf, J. Peterson, and L. H. Jensen. J. Am. Chem. Soc., 80:3835 (1958).
99. N. Ananthakrishnan and R. Srinivasan. Indian J. Pure Appl. Phys., 2:62 (1964).
100. L. K. Steinrauf and L. H. Jensen. Acta Cryst., 9:539 (1956).
101. J. Peterson and L. K. Steinrauf. Acta Cryst., 13:104 (1960).
102. H. L. Yakel, Jr., and E. W. Hughes. Acta Cryst., 7:291 (1954).
103. A. McL. Mathieson. Acta Cryst., 5:332 (1952).
104. A. McL. Mathieson. J. Sci. Instrum., 28:112 (1951).
105. E. Marsh and J. P. Glusker. Acta Cryst., 14:1110 (1964).
106. G. V. Gurskaya and B. K. Vainshtein. Kristallografiya, 8:368 (1963).
107. B. K. Vainshtein and G. V. Gurskaya. Doklady Akad. Nauk SSSR, 156:312 (1964).
108. G. V. Gurskaya. Kristallografiya, 9:839 (1964).
109. R. Srinivasan. Acta Cryst., 9:12 (1956).
110. R. Srinivasan. Current Sci. (India), 27:46 (1958).
111. R. Srinivasan. Proc. Ind. Acad. Sci., A49:340 (1959).
112. R. Srinivasan. Proc. Ind. Acad. Sci., A50:19 (1959).
113. R. A. Pasternak. Acta Cryst., 9:341 (1956).
114. J. Donohue. Nature, 171:258 (1953).
115. J. Donohue, L. R. Lavine, and J. S. Rollett. Acta Cryst., 9:655 (1956).
116. J. Donohue and A. Caron. Acta Cryst., 17:1178 (1964).
117. R. H. Kretsinger and F. A. Cotton. Acta Cryst., 16:651 (1963).
118. M. M. Harding and S. J. Cole. Acta Cryst., 16:643 (1963).
119. B. A. Wright and P. A. Cole. Acta Cryst., 2:129 (1949).
120. A. McL. Mathieson and H. K. Welsh. Acta Cryst., 5:599 (1952).
121. G. V. Sasisekharan. Acta Cryst., 12:941 (1959).
122. R. L. Kayushina. Kristallografiya, 5:944 (1960).
123. B. K. Vainshtein, I. M. Gel'fand, R. L. Kayushina, and Yu. G. Fedorov. Doklady Akad. Nauk SSSR, 153:11 (1963).
124. R. L. Kayushina and B. K. Vainshtein. Kristallografiya, 10:833 (1965).
125. J. Zussman. Acta Cryst., 4:72 (1951).
126. J. Zussman. Acta Cryst., 4:493 (1951).
127. J. Donohue and K. N. Trueblod. Acta Cryst., 5:414, 419 (1952).
128. L. Pauling. The Nature of the Chemical Bond. Ithaca, Cornell University Press, 1940.
129. W. Fuller. J. Phys. Chem., 63:705 (1959).